Postharvest Techniques and Management for Dry Flowers

NIPA® GENX ELECTRONIC RESOURCES & SOLUTIONS P. LTD.
New Delhi-110 034

Postharvest Techniques and Management for Dry Flowers

P. Aruna, T.L. Preethi, V. Ponnuswami,
V. Swaminathan & R. Sankaranarayanan
Horticultural College and Research Institute
(Tamil Nadu Agricultural University)
Periyakulam, Theni District, Tamil Nadu

NIPA® GENX ELECTRONIC RESOURCES & SOLUTIONS P. LTD.
New Delhi-110 034

NIPA® GENX ELECTRONIC
RESOURCES & SOLUTIONS P. LTD.

101,103, Vikas Surya Plaza, CU Block
L.S.C. Market, Pitam Pura, New Delhi-110 034
Ph : +91 11 27341616, 27341717, 27341718
E-mail: newindiapublishingagency@gmail.com
Web: www.nipabooks.com

For customer assistance, please contact
Phone: + 91-11-27 34 17 17
Fax: + 91-11- 27 34 16 16
E-Mail: feedbacks@nipabooks.com

ISBN: 978-81-19254-19-4

Composed and Designed by NIPA®.

Tamil Nadu Agricultural University

P. Murugesa Bhoopathi
Vice-Chancellor

COIMBATORE-641 003
TAMIL NADU, INDIA

Foreword

India is emerging as one of the best in the world flower trade and the floriculture industry is poised for a quantum leap in the near future. Among the various fields of floriculture the dry flower industry holds a strong position, albeit a underexploited one. Dry flowers constitute 70 per cent of Indian floriculture exports. According to APEDA, India's export of dried flowers is 1,536 mt valued at 6871 thousand US dollars. The global dried flower market is estimated at about Rs 2,000 crores annually. At present, the industry is growing at a fast rate of 20 per cent every year. Since India is blessed with a diverse agro climatic zone and abundant manpower, the dry flower industry has enormous potential for development and popularization. This industry also holds promise towards employing rural women. Preserving plant materials in a dried form is not a new idea. It has been considered as an art for hundreds of years. There are many reasons for drying plant materials. One of them is the abundance of available material and the fact that about 80% of flower species can be dried and preserved successfully. A unique characteristic of dried ornamentals is their versatility. They can be arranged into different crafts to suit individual's preferred style, design and utility and is a promising field to be explored.

In this book, the authors have dealt with almost all aspects of the dry flower technology, in detail. This book promises to be a guide for students, teachers and amateurs and for all those interested in pursuing this fine art. I congratulate the authors in their effort to provide all the information required in a pleasing and concise manner.

P. Murugesa Bhoopathi
TNAU, CBE

Preface

India, a traditional grower of wide range of flowers is emerging as one of the powerful forces in the world flower trade and is a leading exporter of dry flowers. The dry flower industry in India is about fifty years old and was introduced by the British. Export of dried flowers and plants from India has a potential of Rs.100 crores per year. The global dried flower market is estimated at a whooping Rs 2,000 crores annually and the industry is growing at a fast rate of 20 per cent every year. The USA is the largest consumer of dried flowers at US$ 2.4 million annually, followed by Germany and UK. Other importing countries are West European countries, Japan, Hong Kong and Singapore. The Netherlands ranks first in export of dried flowers followed by Columbia, Mexico and Israel.

Though, India contributes a miniscule 1.5 per cent to Europe's requirement, dried flowers accounted for 70 per cent of the total floriculture export from India. A mere 15 exporters from the western and southern port towns of India are presently monopolizing this industry. Blessed with a diverse agro climatic zone and abundant manpower, this labour intensive and highly profitable industry is waiting to be tapped. More important is the promise this industry holds in employing rural women and its viability as a cottage industry especially for farm hands.

There are many reasons for drying plant materials and one of them is the abundance of available materials. It is estimated that about 80% of flower species can be dried and preserved successfully. Sophisticated training and expensive equipment are not needed to come up with variety of designs. Unlike fresh flowers that easily lose their marketable value and quality, dried ornamentals offer longer periods of sale if properly preserved, packaged, and handled. Another unique characteristic of dried ornamental is their versatility. They can be arranged into different crafts according to one's preferred style, design, and use.

The dry flower industry is waiting to be explored and needs to be popularized. This book is an attempt to popularize dry flower technology among students of horticulture, agriculture and amateurs.

Authors

Contents

Chapter-1

Introduction

Flowers attract both young and old alike. They are associated with mankind from the dawn of civilization. These have become an integral part of human life in the modern era.

The Indian floriculture industry has been shifting from traditional flowers to cut flowers and dry flowers for export purposes. The liberalized economy has given an impetus to the Indian entrepreneurs for establishing export oriented floriculture units under controlled climatic conditions. About 60,000 hectares, area is under floriculture at present. During nineties, increasing demands for flowers in world market was coupled with production technology available abroad for high quality flowers.

Though flowers are attractive, they remain fresh only for short while. To overcome this problem, everlasting flowers, also known as dry flowers are becoming popular. The beauty and freshness of cut flowers can be retained only for few days even by using post harvest techniques. However, dried flowers and foliage remain fresh for long time. The dried flowers can stimulate not only the eyes but also the sense of smell. This opens new marketing opportunities for small scale entrepreneurs (Anon., 2003 and Handwerker *et al.*, 1990). The naturalized Indian flora comprises of a wide variety of ornamental trees, shrubs, climbers, herbaceous plants etc. These plants are not suitable for use either as cut flower or foliage. But they can be dehydrated to preserve their natural beauty. In addition, a number of commercial cut flowers can be exploited for dry flower industry (Anon., 2003).

After liberalization and EXIM policy, the export of flower products has sharply increased. Dry flowers constitute 70 per cent of India's floriculture exports and the dry flower industry in India is about 50 years old. According to APEDA, India's export of dry flowers and flower buds are 1536 mt valued at 70 lakh US $. India stands fourth in dry flower exports world wide (Anon, 2006). Dried flower products are in very high demand and add an enriched value to the flourishing industry. Various types of dry flower products are hand made paper, lampshades, wall quilt, decorations, books, candle holders etc. The flower arrangements including dried samples of cone, foliage, flowers like rose buds, lilies and other plant materials enhances the beauty of dry flowers adding more export value to this industry. The major consuming countries of dried flowers are USA, Germany and U.K. The major importing countries are West European countries, Japan, Hong Kong and Singapore. Netherlands ranks 1st in export of dried flowers followed by Columbia, Mexico and Israel (Bhattacharjee and De, 2003).

Scope of Dry Flowers in India

1.1 Rich Bio-diversity

The native and naturalized Indian flora comprises of a wide variety of ornamental trees, shrubs climbers, herbaceous plants, fleshy fungi, lichens, mosses and sellaginellas, etc. Many of these ornamental plants with their magnificent array of flowers, foliage, fruits, cones, seeds, roots stems, shoots/twigs, bark, lichens and fleshy fungi have fascinated the tourists, naturalists, environmentalists and amateur gardeners. These plants cannot be retained as cut flower or cut foliage for longer duration and the masses residing in the metropolitan and other large cities who cannot visit countryside or Himalayan region remain deprived of the beauty of these plants. The charms and beauty of these plants however, can be maintained for months to years by employing various techniques of dehydration. In addition to native and naturalized plants, number of cultivated plants especially flowering annuals viz., Amaranthus, button daisy, celosia, dahlia, gomphorina, marigold, pansy, paper flower, salvia, straw flower, statice, bells of Ireland, etc. and other leading flowers like carnation, chrysanthemum, rose and lotus can be grown for dehydration.

1.2 Dehydration Techniques

For drying of flowers and foliage, number of dehydration techniques are practiced which vary according to the suitability of any species and the purpose for which dehydrated material is required. These methods are air-drying, sun drying, solar drying, hot air oven drying, microwave drying, freeze drying and vacuum drying etc. The flowers can be dried or with embedding in desiccants. The important desiccants are silica gel (white and blue), borax acid, white river sand, alum powder, corn powder, saw dust, etc. The flowers and foliage are also press dried in herbarium press, which are required to be changed daily in news paper or bottling paper initially for few days to soak the moisture quickly. The most ideal temperature for flower drying in hot air oven is 45-65°C, which is given for few hours to three days depending upon the flowers species. In microwave oven, plants are normally dried for 2-5 minutes and the material is kept at room temperature for 4-5 hours as settling time. In microwave oven, glass or plastic containers are used whereas in hot air oven, aluminium containers are most ideal. The methods of dehydration techniques have been elaborated in subsequent chapters.

1.3 Floral Crafts

The dried flowers, foliage, fruits, cones seeds, roots, bark, stem/ twigs are used for making various value added floral crafts viz., sweet smelling pot-pourri, collages, flower pitchers, flower balls, greeting cards, pomanders, wall sceneries, festive decorations, bouquets, wreaths and driftwood.

1.4. Future Strategies

We are exporting over ten million pieces of lotus pods per month in addition to dried flowers of camellia, dahlia, bell cups, marigold, Jute flowers, wood roses, wild lilies, paper flower and naturally dried plant parts of Himalayan flora. In-spite of rich bio-diversity and cheap labour in India, no organized production of dehydrated flowers and floral crafts has been undertaken. This business is confined to about 50 private companies operating from Kolkata (West Bengal), Tuticorin (Tamil Nadu) and Mumbai, close to sea ports as the material being bulky is usually shipped through sea routes. Only certain companies have been maintaining a monopoly in the dry flower trade.

At present, the work is in progress under ICAR's adhoc research project on dry flowers at Himachal Pradesh Agricultural University, Palampur to identify potential dried ornamental plants in outer Himalayan region and to standardize their dehydration technology for making best use of Himalayan bio-diversity. This will help in fulfilling the increasing demand for dry flower products in our country and world-over. Dried flowers being cheap and everlasting are fast becoming a favorite among the adventurous flower lovers across the globe. The organized research and production of dry flowers in the country by making use of rich bio-diversity and cheap labour would further trigger the growth of floriculture industry (Anon,. 2003).

Summary

- India stands first in dry flower export.
- Dry flowers don't mean only flower parts, but also includes dried shoots, seeds, pods and bark of trees.
- Export of dried flowers and plants from India is about Rs.150 crore per year. The industry exports 500 varieties of flowers to 20 countries.
- The commonly used dehydration techniques are air-drying, drying by using various desiccants, microwave drying, freeze drying and vacuum drying.
- Some of the important products that can be developed from dry floral materials are pot-pourri, festive decorations, bouquets, wreaths, driftwood and greeting cards. Some other products like wall hangers and flower arrangements can also be made.

Questions and Answers

1. What is the scope and importance of dry flowers?
2. Mention the export potential of the dry flower industry in India.
3. What are the various products from dry flowers?

References

1. Anon. 2003. Science Tech Entrepreneur. Oct .Website: http://www.tibetancc.org
2. Anon. 2006. Website: http://data.un.org

3. Bhattacharjee, S.K. and L.C. De. 2003. Dried flowers In: Advanced commercial floriculture Vol. 1. Aavishkar Publishers, Jaipur. pp 162-173.

4. Handwerker S. Thomas, George B. Olcomb and Howard W. Bud Kere, 1990 Speciality Flowers - A small scale Agriculture alternative Website: http://www.sfc.ucdavis.edu/pubs/brochures/specialityflo.html

Chapter-2

History of Dry Flowers

Preserving flowers dates back to earliest recorded history. The first evidence of use of dry flowers was when the bones of a pre historic man discovered in the Middle East were found with a mantle of delicate wild flowers. The pollen grains that remain indicated specific flowers that were used. During ancient times people used to freshen or enliven their rooms with fresh sweet smelling flowers. In the sixteenth century, when the Elizabethan Age was coming to a close, it was considered fashionable to wear silk flowers as part of one's attire. These days flowers are dried and are versatile in their use. They supply us with food, medicines, ornaments and adornment. Flowers became essential for a fashionable appearance. Each civilization developed its own interpretation of flowers. Pressed flower art became popular in Victorian times. The Japanese use pressed plants and colorful papers to create Oshibana Art.

2.1 Japanese Civilization

For centuries, the Japanese have given special attention to the flowers and plants around them giving special reverence to the elements of the world and all living things. This led the Japanese to creat Oshibana, the art and craft of pressing flowers to retain the benefits and beauty of the living plant. This is also called "pressed flower art." Over the centuries, enthusiasm for this craft has spread from Japan to Victorian England, through Europe, and now, to the United States. Today, Oshibana offers us more than simply retaining the beauty and grace of living flowers. By working with pressed flowers to create enduring designs, we are stimulating the memory of our connection to nature. And when this occurs, the benefits are impressive. With a creative approach to the use of materials, a leaf becomes a tree and petals form mountains. With meticulous attention to detail each leaf and flower is glued onto a precise location.

2.2 Egyptian Dry Flower Industry

Ancient Egyptians preserved elaborate garlands and made detailed preparations using grains and herbs for their dead to enjoy in the next world. Medieval monks harvested and dried flowers and herbs for medicinal purposes. Victorian ladies considered floral garlands to be an essential fashion accessory. They displayed dried flowers in glass domes and designed clever pictures using dismembered cones, lavender, barley and ribbons.

Herb fragrances have long been valued to mask unpleasant odours, create a pleasant environment, evoke memories, make the air fresh, and scent cosmetics, perfumes, and household products. In fact, the use of perfume dates back to ancient Egypt where it enhanced royal and religious ceremonies. Ancient Egyptians preserved elaborate garlands and made detailed preparations using grains and herbs for their dead to enjoy in the next world. Medieval monks harvested and dried flowers and herbs for medicinal purposes. Victorian ladies prepared a range of articles from floral garlands to glass dome and designed clever pictures using dismembered cones, lavender, barley and ribbons. The interest in dried flowers comes in waves bringing all the old applications and techniques along with fresh inspirations and ideas. Today, dried flowers are longlasting, tolerant of most temperatures and offer a wide range of subtle and striking colors

Preserving plant materials in a dried form is not a new idea. It has been considered as an art for hundreds of years. Fragrant dried herbs were placed in Egyptian pyramids. During the middle ages, monks used dry flowers, foliage and herbs for decoration. Dry flower arrangements have been popular in Europe for several centuries. During 1700 A.D, the early American settlers used dry flowers to brighten their homes. Any plant material can be preserved for use as dry flowers (Al Pertuit, 2004).

Summary

- The first evidence of dry flowers was the Middle East.
- Medieval monks harvested and dried flowers and herbs for medicinal purposes.
- Victorian ladies considered floral garlands to be an essential fashion accessory. They displayed dried flowers in glass domes and designed clever pictures using dismembered cones, lavender, barley and ribbons.
- The Japanese use pressed plants and colorful papers to create Oshibana Art. Which is also called "Pressed flower art".
- In the sixteenth century, in the Elizabethan Age dried flowers were used.
- Pressed flower art became popular in Victorian times.
- During 1700 A.D, the early American settlers used dry flowers to brighten their homes.

Questions and Answers

1. Write a brief history of the dry flower industry.
2. What does the Oshibana art indicate?

Reference

1. Al Pertuit. 2004. Drying Flowers. Website: http://hgic.clemson.edu/

Chapter-3

Dry Flower Industry

3.1 Indian Floriculture Industry

Indian floriculture industry has been shifting from traditional flowers to cut flowers and dry flowers for export purposes. The liberalized economy has given an impetus to the Indian entrepreneurs for establishing export oriented floriculture units under controlled climatic conditions. About 60,000 ha area is under floriculture at present. During nineties, increasing demands for flowers in world market was coupled with production technology available abroad for high quality flowers. This encouraged some entrepreneurs to establish export-oriented units in Hyderabad, especially for the production of long stem rose by importing cultivars and adapting green house technology.

3.1.1 Indian Dry Flower Market

The most promising area in floriculture is the dry flower industry. Dried flowers and plants have been exported for the last 40 years and today, India is one of the leading countries in the field. Export of dried flowers and plants from India is more than Rs 150 crores per year. The industry exports 500 varieties of flowers to 20 countries and dry flowers are highly in demand in USA and UK markets. The potpourri variety of dried flowers is a purely export oriented and has specialized processing units.

India, with its vast resources, varied products and experience in the field of dried flowers and plants enjoy a distinct advantage. In Andhra Pradesh, there is abundant availability of raw material required for dry flower making.

Dried flower products are in very high demand and add an enriched value to the flourishing industry. Various types of dry flower products are handmade paper, lampshades, wall quilt, decorations, books, candle holders, etc. The flower arrangements using dried samples of cone, foliage, flowers like rose buds, lilies and other such plant material enhances the beauty of dry flowers, adding more export value to this industry.

3.2 Australian Dry Flower Industry

Australia is now an integral part of the expanding world flower trade, which in the year 2000 was worth an estimated US $7 billion. Australia's potential for exporting flowers lies in its unique native flora. The true value of the Australian export industry is not clearly defined but is likely to be around 50 million dollar per annum. Australian flowers account for around 95%. Domestically and internationally, the industry is recognized for its strict production standards, a sophisticated supply chain and high quality products.

3.3 NSW Dry Flower Industry

The New South Wales flower industry has an estimated production value of 202.7 million dollar, with most produce selling on the domestic market. Native and South African flowers account for about 10-15% of total domestic sales.

3.4 Major Producers of Dry Flowers

Flowers are grown on a global scale. The biggest flower growing nations are the Netherlands, Colombia, Kenya and Israel, but many other countries like India, South Africa, Ecuador and Malaysia are now investing in the industry. These flowers are transported to several countries where they are dried and used as dried flowers.

In India, Maharashtra, Karnataka, Andhra Pradesh, Hyderabad, and Kolkatta have emerged as major dry flowers industries in recent times. The demand for dry flowers are also met through production at coastal areas of Nalgonda, Medak, and in the West Godavari districts of Andhra Pradesh. The other countries having a name in the dry flower industry are Europe, Japan, Taiwan, UK, Hongkong, and Australia.

Summary

- Dried flowers and plants have been exported for the last 40 years and today, India is one of the leading countries in the field. Export of dried flowers and plants from India is more than Rs 150 crores per year.
- Australia is now an integral part of the expanding world dry flower trade, which in the year 2000 was worth an estimated US $7 billion.
- The Indian dry flower industry exports 500 varieties of flowers to 20 countries and dry flowers are highly in demand in USA and UK markets.
- The New South Wales flower industry has an estimated production value of 202.7 million dollar, with most produce selling on the domestic market. Native and South African flowers account for about 10-15% of total domestic sales.

Questions and Answers

1. What are the major dry flowers producing countries?
2. What is the annual dry flower production of India?
3. How much percentage of dry flower export contributes to the overall floriculture exports of India?

Chapter-4

Harvesting of Dry Flowers

4.1 Dry Flower Garden

Contrary to popular assumptions, the area required for producing large quantity of dried flower material need not also be large. However, the garden should be planned so as to harvest dry flowers all year round by planting evergreens, deciduous plants, annuals and perennials. Rows of flowering plants should be planted in the dry flower garden. Herbs are easy to grow, and can be planted in pots and window boxes. They are useful in transforming simple cooking into gourmet dishes, while also lending their delicate scents to a room when dried and for use in wreaths, pot pourri and arrangements. Gardens for dry flower production should include a wide selection of weeds, grains, seed-heads and pods. Flowers such as hydrangeas, zinnias, roses, dahlias and sunflowers and a selection of line flowers such as golden rods, statice, gladiolus, larkspur and delphinium can also be grown as these will provide a greater variety of flowers needed in flower arrangements.

4.1.1 Harvesting Flowers

Flowers and other plant materials for drying should be picked close to their prime. Flowers or leaves for drying may be collected at any time during the growing season. The best time of day to cut is midmorning, after the dew has dried but well before any flowers wilt. Dampness can lead to mould and slows drying. During rainy weather, plants should be harvested when they are dry even if it means cutting them a little early. Damp or over matured plants should not be harvested. If harvesting flowers when wet is unavoidable, the excess moisture should be gently blotted off with a soft paper towel or tissue.

Flowers to be air-dried continue to open as they dry; therefore such flowers should not be fully open at picking. Wilted flowers should never be placed into drying mixtures. More material than is needed should be always collected, to allow for damage. Only the most perfect forms should be used. Poor shapes dry as poor shapes. Plants and flowers free of damage due to pest and disease incidence should be selected. This is because any such damage becomes only more obvious after drying. There are definite developmental times, which are best for cutting flowers for drying. This can be very specific for different plants or even different cultivars of the same plant. In case of cut roses, flowers should be harvested just before they reach full bloom and hung up to dry. Delphiniums, larkspur, foxglove and lupin should be gathered when the lower buds are flowering, but the top ones are still closed. Visible pollen seeds and wilting or missing florets on flower spikes indicate an over-mature flower. In general, flowers should be cut when the heads feel firm, just before they come into full bloom.

It is sometimes difficult to develop graceful lines when making dried flower arrangements. Therefore, while collecting material for dry flowers, branches and stems with sweeping curves or lines that will add distinctiveness to the arrangement should be selected. If none can be found, curves or other lines can be made by shaping the branches or stems into the desired positions while they dry.

Stems should be promptly placed in a container of water to prevent wilting while gathering. Once cut, stems should be grouped into bunches using rubber bands and removed from the sunlight as soon as possible. All the flowers should be dried with their natural stems. However, natural stems do not give the flower enough support when dry and may be difficult to manipulate and position. So, most of the stem is cut, leaving only about an inch. Stems can be replaced with florist's wire after drying. Flowers should be wired before drying them because drying may cause shattering in some cases.

4.1.2 Harvesting Foliage

Harvesting foliage in their active phase is preferable because at this stage, the leaves will be at peak maturity. Foliage is dried by preserving in a solution of water and glycerin. However, certain important points have to be borne in mind while harvesting foliage for preserving with glycerin. Young green leaves and leaves in dormant stage should not be harvested because young green leaves will not absorb the glycerin solution and the leaves in dormant stage would have stopped drinking water. So, both are unsuitable for absorbing glycerin.

4.2 Pre Drying Flowers

Flowers that dry well are typically colorful, compact, strong-stemmed and relatively low in moisture content. Before the flowers are put into the drying process there are several measures to be taken for successful drying of the flowers.

a) *Cleaning* : After the plants are harvested or gathered, they must be cleaned. Cleaning may involve screening, washing, peeling, or stripping leaves from stems. Any unnecessary parts are removed prior to drying to avoid wasting time and energy. Cleaning is often done by hand.

b) *Stem Supports* : Many flower heads are too heavy for the dried stems to support during the drying process. The stems have to be cut and wired before hanging them out to dry. The wired stems have to be covered with floral tape after the flowers are dry. Some flowers can be wired after they are dried, but it is easier to wire them when they are fresh, because the wire slips through the moist heads more easily. Several stems of one type of flower or seed pod are wired together for a denser, brightly colored effect. If a flower head falls off during the drying process, a floral stem wire should be hot glued to the head or calyx of the dried flower, and the wired stem covered with florist tape.

c) *Adding Wire Stems* : Each flower head from the spray has to be cut so that the stem is approximately 1½ long. Using 20-gauge wire, the wire should be cut to the desired length, allowing extra for bending and for the part of the stem to be inserted into the floral foam. The tip of the wire with pliers has to be bent to form a 'shepherd's hook'. The wire should be hooked around the calyx, where the flower head joins the stem, and squeezed tightly with needle nose pliers.

Summary

- Dry flower gardens should be planned so as to harvest dry flowers all year round by planting evergreens, deciduous plants, annuals and perennials.
- Herbs are easy to grow, and can be planted in pots and window boxes.
- Flowers or leaves for drying may be collected at any time during the growing season. The best time of day to cut is midmorning, after the dew has dried but well before any flowers wilt.
- Only the most perfect forms should be used.
- Harvesting foliage in their active phase is preferable because at this stage, the leaves will be at peak maturity. Foliage is dried by preserving in a solution of water and glycerin.
- Before starting the drying process there are several steps to be followed for successful drying of the flowers. They include cleaning, providing stem supports and adding wire stems.

Questions

1. Write a short note on planning a dry flower garden.
2. What are the points to be considered while harvesting dried flower materials?

Chapter-5

Natural and Air Drying

Various methods are employed to dry flowers and foliage. The method adopted for drying depends largely on the character of the plant, that is to say whether the plant is glabrous, succulent or not succulent, hardy or delicate in nature. The suitable method for drying any given plant has to be studied elaborately. However, the different methods of drying are described broadly in this and consequent chapters.

5.1 Natural Drying

Natural drying is a process that has been used since ancient times. It involves sun drying in the field. It is the easiest method for drying flowers such as delphiniums, globe thistle, celosia, and hydrangea. This method requires no drying equipment. It uses solar energy and requires large amount of space. The plants can be damaged by exposure to adverse weather conditions. The exact length of time for adequate drying depends on temperature and humidity. Usually, a little grooming is all that is necessary for naturally dry materials. However, cones and pods may need to be washed in water and a mild detergent. Fragile seed heads, such as those of pampas grass, as well as mature cattails, can be sprayed with hair spray or other aerosol lacquers or plastics to hold the heads in place and to prevent shattering as the plants age. Lacquers or shellac can be sprayed or painted on fruits and cones to give them a shinier, decorative look as well as to help preserve them. Seeds from pine cones are removed to prevent shedding that may occur at a later time.

5.2 Air Drying

Air drying flowers is one of the easiest methods of preservation and gives plants a crisp look that lasts for years. Air drying flowers make a fabulous decoration by themselves, but when they are dry, they make more beautiful and exotic flower arrangements. This is commonly referred to as the "hang and dry" method. It is the oldest and easiest drying technique. No special equipment is needed. The stems of flowers and their foliage are tied and hung upside down. The rooms should be warm, dark and dry with good air circulation. However, it is one of the longest drying methods. It usually takes three to four weeks for the flowers to dry completely. The flowers are hanged upside down so that the stem remains straight. If they are hanged with right side up, they would bend over and the result will be dried flowers with distorted stems.

The ideal place for hanging the flowers is a place with ideal conditions like:

- Darkness
- Very good airflow
- Perfect humidity levels

Flowers are then sprayed with hair spray or clear varnish to retain their form. Air dried products tend to loose their original colour faster (Fresco, 2002).

Ornamental grasses, cat tails and branches dry well by upright method. The leaves of some trees and shrubs like oak cuttings should be placed flat in a shallow box lined with sheets of blotting paper. Soft flat topped flower heads such as Queen Ann's lace, dry best when their stems were inserted in wire mesh. The drying process usually takes 10 to 20 days, depending on the plant, for shade drying. The test for dryness is commonly done by dissecting one or two flowers. Annuals like amaranthus, dusty miller, nigella and statice and perennials like artemesia, baby's breath, chinese lantern, lamb's ear, lady's mantle, oregano and yarrow were quite suitable for natural drying.

5.2.1 Steps of Air Drying Process

- Flowers of good quality are cut at prime condition or when slightly immature.
- Foliage is removed from stems. If stems are weak or become brittle after drying, they are removed and flowers are wired.
- Stems are grouped into small bunches and tied with a rubber band. This will pull stems tighter as the stems shrink during drying.
- Stems are hung upside down in a warm, dry, dark area such as an attic, closet or furnace room. Damp rooms or direct sun on the flowers should be avoided. Good air circulation is important.
- Flowers should be dried completely and plants are sorted by flower types and sizes, and stored in an airtight container until ready to use.

5.2.1.1 Steps for Air Drying of Rose Flowers

Rose buds should be harvested at tight bud stage and allowed to open at least halfway before the drying process is initiated. Work area should be cleared of rough materials such that the delicate petals are not damaged. The end of the stem has to be bound to the hanger, so that the flower can be oriented naturally downwards. For drying multiple flowers, they should be arranged along the hanger such that it will be balanced; conversely, if only one stem is to be dried, the stem should be hanged in the middle of the hanger. It is important to hold the flower out of contact with anything in the work area while tying in the hanger, otherwise the petals could be greatly weakened. The time taken for the flowers to dehydrate largely depends on the humidity of the drying area, but three weeks is usually sufficient. It is critical not to touch the flowers during this time. The flower's color will darken as it dries. A gentle layer of hair spray has to be sprayed to the flower. It will greatly improve the strength of the dried flower. A second coat can also be applied, if desired.

5.2.1.2 Flowers for Air Drying

Picking the flowers at the correct stage of development for air drying is very important for the success of the drying process Flowers such as Hydrangea, Gypsophila, and *Alchemilla mollis* should be picked, then the stems placed in a vase with about an inch of water in the bottom. By the time the flowers have used all the water, they will have dried successfully.

Hydrangeas need to be turning "papery" to the touch before drying, otherwise they will just shrivel. Most other flowers should be well developed before drying. Flowers with a papery feel (often known as "immortelles") usually dry very well.

Helichrysums should be picked when the outer two or three rings of petals have developed. If the centre of the flower is on view, then it is too late to pick and dry them successfully. This is because Helichrysums carry on developing for a while after they have been picked, and will open out backwards on themselves, eventually shedding all the petals when dried.

Plants suitable for hanging include those that are papery and somewhat dry and stiff even in their natural form. Some of the plants are cockscomb, lavender, statice, strawflower, yarrow, herbs, thistle, sage, roses, marigold, iris, and larkspur.

Blue and yellow flowers retain their colors when air dried, but pink flowers fade. Rose flowers and peonies shrink somewhat when air dried. Annuals like zinnias and marigolds are easily dried. Everlasting flowers like strawflower, globe amaranth, cockscomb, and baby's breath are also suitable.

Some flowers that are suitable for air-drying are: Yarrow (*Achillea millefolium*), pompon Dahlias (*Dahlia hortensis*), Poppy seed heads (*Papaver somniferum*), Roses (*Rosa spp*), Marjoram (*Origanum vulgare*), Delphinium, Larkspur (*Consolida ambigua*), Lavender (*Lavandula Augustifolia*), African Marigold (*Tagetes erecta*), Strawflower (*Helichrysum bracteatum*), Globe Thistle (*Echinops ritro*), Cornflower (*Centaurea cyanus*), Statice (*Limonium sinuatum*), Globe amaranth (*Gomphrena globosa*), and Love-in-a-mist (*Nigella damascena*) seed heads.

5.3 Water Drying

Water drying is a method of preserving in which the leaves are stripped off and the flower stems placed in five centimeters of water. This is then, placed in a warm place, out of direct sunlight. The water is absorbed and evaporates as the flower dries. Hydrangeas, heathers, hybrid delphiniums, acacia, gypsophila, bells of Ireland, proteas and yarrow dry well this way. The best way to dry hydrangea flowers is by water drying. The freshly cut flowers are placed in a vase with fresh water. The stems are at least half covered with water. The flowers will still look attractive. More water should not be added as the water in the vase evaporates. The water allows hydrangeas to dry naturally, rather than simply dry out. Once the water is totally evaporated, hydrangeas should feel dry and will then be ready for use. Bells of Ireland, delphinium, gypsophila and hydrangea best retained their shape when placed in about five centimeters of water (Anon., 1997a). Al Pertuit, (2004) and Arul Murugan *et al.*, (2007) reported that some flowers dry well if placed in water.

Summary

Air drying flowers is one of the easiest methods of preservation and is commonly referred as the "hang and dry" method. The stems of flowers and their foliage are tied and hung upside down. The rooms should be warm, dark and dry with good air circulation. However, it is one of the longest drying methods. It usually takes three to four weeks for the flowers to dry completely. There are various steps involved in air drying. Flowers such Dahlias (*Dahlia hortensis*), Roses (*Rosa spp*), Delphinium, Larkspur (*Consolida ambigua*), African Marigold (*Tagetes erecta*), Strawflower (*Helichrysum bracteatum*), Statice (*Limonium sinuatum*) and Globe amaranth (*Gomphrena globosa*) are suitable for air drying. Water drying involves removing the leaves and placing the flower stems in five centimeters of water. This should be placed in a warm place, out of direct sunlight. The water is absorbed and evaporates as the flower dries. Hydrangeas, heathers, hybrid delphiniums, acacia, gypsophila, bells of Ireland, proteas and yarrow dry well this way.

Question and Answers

1. What is air drying? Mention the steps involved in air drying.
2. Describe the air drying method for roses.
3. Mention flowers suitable for air drying.
4. What is water drying? Mention some flowers suitable for this method of drying.

References

1. Al Pertuit . 2004. Drying Flowers. Website: http://hgic.clemson.edu/
2. Anon. 1997. Is it real or freeze dried? Website: http://w3.goodnews.net/~aretz/dry.htm.

3. Arul Murugan, P., G. Thiyagarajan and K. Ramesh. 2007. Dry flower technology. Science Tech Entrepreneur. Website: http://www.tibetancc.org/Dried%20Flower.htm

4. Fresco, Mary Charlotte, 2002. The fresh business of dried ornamentals. Archives: Crops: (4):2 Website: http://www.bar.gov.p/bardigest/2002

Chapter-6

Press Drying

Pressing Flowers

One of the most popular methods for drying flowers is to put them under pressure, to remove the moisture out while leaving the color of the flowers and structure intact. There are several ways to apply pressure to flowers. The easiest method is placing them in heavy books and allow for drying. Flower presses can also be used. Some guidelines for pressing flowers in the traditional way are explained below:

Pressing is a very easy way to preserve flowers. The flowers will become flat. Unglazed paper, such as newsprint or an old telephone book, is best for pressing. The flowers are spread so that they do not overlap and are placed between newspaper sheets. Additional layers of paper and flowers can be built up and then covered with a board or piece of cardboard before pressing down with a heavy object. The time required for drying, depends on the flower size or tissue content and can be anywhere from two to four weeks. The process can be speeded up by placing a stack of papers and flowers over a light bulb.

Storing pressed flowers is not a problem because they usually are not removed until they are used. Succulent leaves tend to just squash when placed in the press. Odd shaped flowers such as daffodils need to be cut into halves and opened out before pressing, and thick flowers such as Chrysanthemums need to have the calyx reduced in thickness. Single petals can also be used, and reassembled when making picture. The flowers and foliage should be cut to press when the weather is dry. Any water trapped in the flowers before pressing will turn them mouldy. Trinklein (2006) reported that pressed flowers are especially suitable for flower pictures, greeting cards etc. Flowers that are in prime condition are used for pressing. Also, flowers with different stages of development up to full maturity are used for more variety in design. Plants with fleshy stems and leaves, flowers with very thin petals and wilted materials are not pressed. Flowers that are flat, such as pansies, are best for press drying. The faster the flowers dry, the better they retain color. On the other hand, flowers can't be exposed to excessively high temperatures; although they may dry quickly, they will turn brown.

Pressing requires sandwiching flowers and foliage between layers of an absorbent material. This should be clean and hold the flowers firmly and flat during the drying process. Porous materials that allow some air movement are also beneficial. Flowers are generally placed between sheets of a non glossy type of paper. Newspapers, old telephone directories or catalogs are suitable. Absorbent facial tissues placed on the pages aid rapid moisture absorption. Tissues should be removed and flowers or foliage replaced between fresh, dry tissues and papers at the end of the first week. After the flowers and tissues have been placed in the folded newspapers or books, they are stacked several layers deep. Boards are placed beneath and on top of the stack. The stack is placed in a warm, dry place with a heavy weight on top.

Another satisfactory system suitable for drying flowers uses a combination of cardboard, newspaper and desk-sized blotter pads. Corrugated cardboard is cut into sheets slightly larger than the sheets of folded newspaper. Flowers are positioned on one side of the opened newspaper. Then, the newspaper is closed and a sheet of blotter paper is placed on either side. After all the flowers have been placed between the newspapers, blotter paper and cardboard, the layers are stacked and tied or taped together. They should then be placed in a warm, well-ventilated place. A heavy weight is placed over this. Special presses can be purchased or constructed for drying large quantities of materials.

Pressed flowers are especially suitable for flower pictures, as well as decoration on note paper, place cards and many other items. Most foliage and simple flowers with few petals press very well. Pansies are an excellent flower for pressing at all stages of flower development. Ferns make excellent pressed plants. Leaves and branches with foliage can be pressed to form plant materials with natural curve. In this process, the faster the flowers dry, the better they retain color. Flowers should not be exposed to excessively high temperatures because high temperature may turn the flowers brown.

6.1 Collection for Pressing

Flowers that are in prime conditions are collected. Flowers with different stages of development up to full maturity can also be used to get more variety in design. Plants with fleshy stems and leaves as well as flowers with very thin petals are avoided. Wilted materials are not pressed and flowers that are flat press best. The fresher the flowers, the better they will press. Whether the flowers are pressed immediately or after a few days it should be borne in mind that they are in good condition. Any wilted petals or curled leaves should be removed.

Any stamens should be carefully removed with pollen that may stain flower flaws or bruises on flowers will become more obvious after drying. This can add to the individual charm of pressed flowers and be quite unique, but to create perfectly preserved flowers, a flower with perfect bloom should be selected. Thinner flowers dry faster, and the faster a flower dries, the better its color will be preserved.

Method

Flowers and foliage are sandwiched between layers of an absorbent material. This should be clean and hold the flowers firmly. Weight or pressure should be applied for at least 5 to 10 days or until the plants are dried and placed in a warm, well-ventilated place. After a week, the absorbent material should be checked for excess moisture, and the material has to be replaced if necessary. The flowers should be repositioned if needed. After drying, storage should be done in an airtight container.

6.2.1 Absorbent Materials

Flowers are generally placed between non-glossy types of paper. Newspaper, old telephone directories or catalogs are suitable. After placing the flowers in the folded books, the books should be stacked several layers deep. Boards should be placed beneath and on top of the stack. The stack should be placed in a warm, dry place with a heavy weight on top.

Another material suitable for pressing is cardboard. This forms a system including cardboard, newspaper and blotter pads. Corrugated cardboard is cut into sheets slightly larger than the sheets of folded newspaper. Flowers are positioned on one side of opened newspaper. Then the newspaper is closed and a sheet of blotter paper is placed on either side. Then the layers are stacked and tied or taped together and placed in a warm ventilated place.

6.2.2 Pressing Flowers in Books

The flowers are placed in a heavy book and allowed to dry, but this method takes at least a month. An old phone book is a good choice, or a thick encyclopedia. Placing flowers between a couple sheets of acid-free blotting paper or even printer paper, which will help absorb moisture, protect the book, and keep the petals from absorbing ink or being degraded by residual printing acids. Application of any additional pressure, such as stacking books or other heavy objects on top, will help speed the process.

The flowers should be checked frequently, and gently straightened for any furled leaves or petals. If the surrounding paper is damp, the flowers should be carefully peeled off by bending the page away and then pressing is continued in the book with dry paper. The flowers are ready when they feel dry to the touch and are rigid, not limp. This method works best with simple garden flowers and petals, but can make thicker flowers look squashed.

6.2.3 Pressing Flowers in a Flower Press

There are a variety of flower presses available for sale both online and in local craft supply stores. A range of sizes and designs are available, and the results from these presses are usually quite good. For example, flower presses that use Velcro straps or rubber ties to create pressure are available. Microwave presses are also available. Flower presses are usually simple to use and come with easy to follow directions, but are costly.

6.2.4 Making A Flower Press

Creating a flower press is easy and requires very little investment. Two pieces of corrugated cardboard, newspaper, a few sheets of blotter or printer paper, and three large rubber bands are required to use this press. The flowers should be placed between a few pieces of paper, between the pages of the newspaper, the newspaper is sandwiched between the cardboard, and held together with the rubber bands, evenly spaced. Pressure is applied by placing heavy objects on top. It takes up to a month or more to dry thick flowers completely, but thinner flowers may be done in about two weeks. To prepare a flower press of higher quality the following materials are needed:

- 2 wood boards of equal size and thickness
- 1 inch thick piece of polyester fiberfill
- 2 pieces of fine nylon fabric
- 1 can of spray fabric adhesive
- 4 bolts, an inch longer than the total thickness of boards and polyester fiberfill with washers and wing nuts.
- Drill with drill bit size appropriate for bolts.
- Clamps

For building the press the boards should be placed on top of each other, making sure the edges match up, and then clamped securely together. Appropriately sized drill bit should be used to make 4 holes, one in each of the corners, about an inch from the corner point. Two boards with four holes each that will match up perfectly when the bolts are inserted should be made.

After drilling, the boards should be separated and the top and bottom of the press should be decided. The best looking board should be kept at the top, which can be decorated after building the press. The bottom board should be taken and should be sprayed with fabric adhesive, then placed fiberfill down in the center. The top of fiberfill should be sprayed with the adhesive, and covered with the first piece of nylon fabric. The nylon and fiberfill should be pressed down to make sure the adhesive will stick. Next, the adhesive should be sprayed on the inside of the top board and second piece of nylon should be pressed into it, doing the best to make sure that it will match up with the fabric on the bottom board. This is then allowed to dry.

The final step in building the flower press is to simply insert the bolts from the outside of bottom board through to the outside of the top board, and screw on the wing nuts. Thus the flower press can be prepared and decorated with paints or with our own pressed flowers

6.3 Flowers for Press Drying

The various flowers recommended for pressing are:

Ageratum
Johnny Jump-Up, Viola
Celosia, Cockscomb
Daffodil
Bleeding Heart
Chrysanthemum
Cornflower, Bachelor's button
Primula
Lily- of the Valley
Queen Anne's Lace
Sweet Pea
Nemesia

Summary

Pressing is a very easy way to preserve flowers. One of the most popular methods for drying flowers is to put them under pressure, to basically squeeze the moisture out while leaving the flower's color and structure largely intact. Pressing requires sandwiching flowers and foliage between layers of an absorbent material. The faster the flowers dry, the better they retain color. On the other hand, flowers can't be exposed to excessively high temperatures; although they may dry quickly, they will turn brown. Pressed flowers are especially suitable for flower pictures, as well as decoration on note paper, place cards and many other items.

Question and Answers

1. Describe the procedure for press drying.
2. Mention the flowers suitable for press drying.

Reference

1. Trinklein, David. 2006 Storing dried material. extension. missouri. edu. Website: http://www.muextension.missouri.edu/explore/agguides/hort/go6540.html

Chapter-7

Drying by Embedding

Drying by Embedding

A desiccant is simply a substance with a high affinity for water which can be used as a drying agent. Embedding the flowers in a granular, desiccating material is probably the most commonly used method and many consider it the best all around method. Several materials are used as drying agents. All the agents used vary in cost and in the results they produce. It is important to use the correct procedure when covering the flowers so that their form is maintained. Most well known is silica gel, but also clean dry sand can be used. Usually an airtight container is used; the flower heads are placed in the drying mixture face up, and very carefully covered with the mixture. The container must be kept closed during the drying process. After 4-14 days, depending on the thickness of the flower, the flower will be dry.

7.1 Embedding Techniques

About an inch of desiccating material is put at the bottom of the container. The flower stems are cut to about half an inch and inserted into the center of the material. The desiccating material is poured along the perimeter of the container. Then it is tapped lightly on the container so that the material will move to the flower, without altering the form of petals. In the same way the material is continuously added until the flower is completely covered. The materials such as sand, silica gel, glycerin and some home made agents can also be used. Before embedding, flowers should be wired as shown in fig (1).

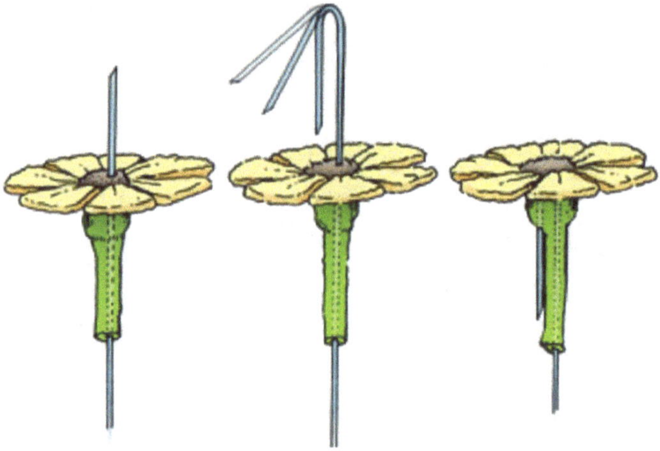

Hide the hook well when hook wiring a flower before drying.

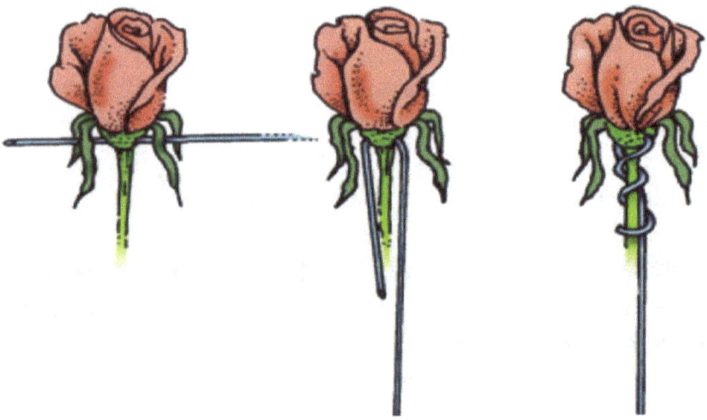

Cross wiring flowers with a hard base like roses

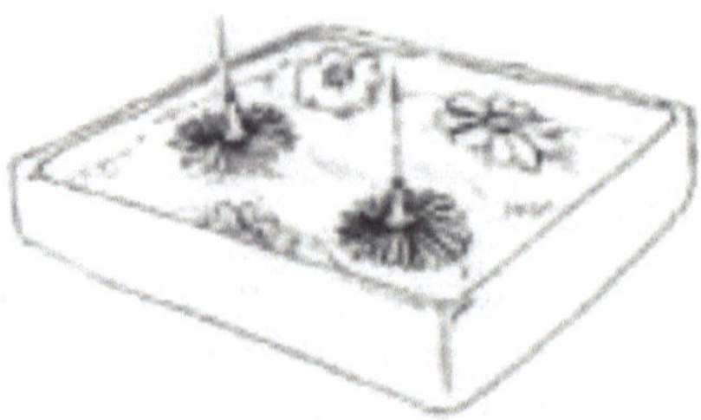

Fig. 1 :

7.1.1 Drying in Sand

Sand must be very fine, clean, dry and preferably salt free. Sieving the sand is recommended to remove coarse grains and foreign particles. Rinsing the sand in water several times to remove any soil is also recommended. Damp sand can be dried in an oven by placing in shallow pans and baking at 120^{0}C for 20 to 30 minutes. Flowers in their prime should only be used and they should be processed quickly to prevent wilting.

To dry with sand, an inch or two of sand has to be placed in a container and a small amount of sand has to be scooped to form a depression. The flower head should be placed upright in this depression and the sand is pressed in and around the outside of the flower to support it. Next a little sand has to be scooped into one's hand and allowed to trickle in a fine stream around each petal. This should be done starting with the outer petals and worked inward row by row, allowing the sand to build up equally on all sides of each petal so its position and shape are not altered. Flowers dried with sand are fragile so should be very careful when removing them from the sand and stored in a strong carton to protect the petals from breaking.

It takes flowers and foliage about three weeks to dry in sand. To prepare tall, spiky flowers like delphinium, an inch of sand has to be poured in a large flat container. The sand has to be poured in a circular motion around the flower. To prepare round, compact flowers like zinnias, the stems are wired, and placed upside down in an inch of sand. The sand should be build around the edges of the flower. The sand should not be poured on top. Children's play sand can be used since it is washed and screened, and it doesn't contain any chemicals or additives.

The sandboxes can be placed in the sun, the attic or any other warm, dry location. Flowers can be left indefinitely in sand, because sand is inert and won't affect the flower. To speed up the drying, the sandbox can be placed in the oven at 38°C. Borax detergent combined with cornmeal or sand is an expensive material for drying flowers. This mixture is used for flowers that are less stiff.

7.1.2 Drying in Silica Gel

Silica gel can be found in most garden centers, nurseries, florist, or hobby shops. It absorbs moisture from flowers rapidly, thus preserving flower color better than other drying methods. Most flowers will dry in 36 to 48 hours. Silica gel should be used in an airtight container or it will absorb moisture from the air, not from the flower tissues. The silica gel can be re-dried in a warm oven (not in a microwave) and reused for other flowers later. Flowers that dry well in either borax or silica gel include rose, aster, carnation, marigold, dahlia, larkspur, geranium, zinnia, chrysanthemum and delphinium. One word of caution when using the homemade agents or silica gel is that the flowers will sometimes reabsorb moisture and wilt. For best success with flowers dried in an agent, flowers should be displayed in a closed container to keep out dust and high humidity.

While using silica gel or other desiccant to dry flowers, all of the stems, are cut off leaving only a small portion attached to each blossom. Drying whole stems requires a large amount of desiccant and a large container to position long stems or large flower heads like sunflowers correctly. In that case, the flowers should be air dried and then a wire has to be attached on the stem with florists tape to camouflage the wire support.

7.1.3 Homemade Agents

Various mixtures can be made with ingredients found in most kitchens. For example, equal proportions of powdered pumice and yellow corn meal or equal proportions of borax and yellow corn meal. To each portion of either of the mixtures, non-iodized salt should be added. Other grain cereals can be also be substituted for corn meal. These mixtures are usually heavier than sand or silica gel, but they work well. The mixture should be applied like the sand but in an open container and placed in a warm, dry location for about two weeks.

7.2 Glycerin Drying

In this method moisture in a flower or foliage is replaced with glycerin and water. The flower is preserved and not dried. Dried materials (whole bunches or single leaves) retain their natural shape and flexibility. They last indefinitely and can be dusted or even wiped with a damp cloth without risk owing to the leathery texture of leaves. Glycerin drying involves the following process. The bark of the stem is removed and the cut end is smashed using a hammer. This portion of the stem is immersed in the glycerin and water solution. The leaf turns brown with glossy appearance after a month. Many types of foliage can be successfully preserved to maintain softness and flexibility. This is done by immersing leaves or placing stems in a 33 per cent glycerol solution.

Summary

Embedding the flowers in a granular, desiccating material is probably the most commonly used method and many consider it the best all around method. Several materials are used as drying agents. All the agents used vary in cost and in the results they produce. Most well known is silica gel, but also clean dry sand can be used. Sand must be very fine, clean, dry, and preferably salt free. The sand should be sieved to remove coarse grains and foreign particles. Silica gel can be found in most garden centers, nurseries, florist, or hobby shops. It absorbs moisture from flowers rapidly, thus preserving flower color better than other drying methods. Most flowers will dry in 36 to 48 hours. When using silica gel the flowers will sometimes reabsorb moisture and wilt. To avoid re absorption, the flowers should be displayed in a closed container to keep out dust and high humidity. Glycerin is also a good preserving agent, especially for foliage. In this method, moisture in a flower or foliage is replaced with glycerin and water. The flower is thus preserved and not dried.

Question and Answer

1. What is embedding?
2. Describe the desiccants used for embedding process.
3. Describe the procedure involved in drying by embedding method.

Chapter-8

Drying in Microwave Oven

Microwave Oven Drying

Microwave drying is quick and relatively simple. It takes only a few minutes and provides dried flowers that look fresher and more colorful than obtained by other methods. Flowers with thick petals like magnolia are not suitable for drying in microwave. Since flowers vary in moisture content, texture and density, care should be taken to use the same sized flowers from one species at a time. It has been found that many flowers held almost true to life color and form using this process. Brightly colored flowers are best to dry. Flowers such as lilies, roses, violets, zinnias, and dahlias work well with this process.

8.1 Precautions for Microwave Oven Drying

- Start with a low setting
- Set the oven for a few seconds at a time.
- Different settings will produce different colour variations.
- Microwave friendly container should be always used.
- Wired flowers should not be used in a microwave oven

8.2 Procedure for Drying

Sheets of paper towel should be placed on a flat plate with a sheet of blotter paper on top, the flower should be placed on the blotter paper and covered with another sheet of blotter paper, more paper towels should be added on top, and then capped with another plate. This drying should be initiated with one flower at time, but once the timing is set, microwaving can be done on multiple flowers as long as they don't overlap.

Drying multiple flowers takes longer time and this can degrade their color. The low-medium setting or a power range of 2-3, should be used and the material heated for 1 to 2 minutes. Usually, this method will take less than five minutes. Drying is complete when the flowers are rigid and feel dry to the touch. The oven is switched off. However the flowers should be left for a few more minutes to complete the drying process. Flowers like marigolds, chrysanthemums, cornflowers and zinnias dry well in a fan-assisted, convection oven. Non-ventilated ovens are not appropriate; because they generate too much moisture. The material must be dried at a very low temperature for more hours. The flowers are placed in slots in a wire mesh rack allowing the stems to dangle below. The time required depends upon the density of the flowers. Care should be taken to see that the oven doesn't get too hot.

8.3 Microwave Oven Drying Using Silica Gel

Silica gel can be used in the microwave with fantastic results. An inch of silica gel should be placed in the bottom of a small cardboard box, a mound is created in the center to support the flower, the flower is placed face down on the mound, and carefully the box is filled with the rest of the gel. At least an inch should be left at the top of the box to catch any silica gel that might escape during heating and then dried in the microwave on low to medium for one minute, and then the container is left for about half an hour before carefully checking the petals for dryness. If necessary, the drying process will be repeated until the flower is rigid but not brittle.

8.3.1 Method

The container is filled partially with silica gel and the flower is placed in container stem side down. The silica gel is slowly poured around the flower until it is covered. The container is placed in the oven along with a cup of water in the rear left corner of the microwave and the flowers are rearranged. Heating is done on full power. Large flowers take longer time to heat. After heating, the flower must stand in the silica gel until the silica gel is cool. Any flower should remain in the microwave for a minimum of two hours and a maximum of four to six hours. If the flower is removed too soon, it will be warm and limp, and will not be able to hold its form.

List of plants suitable for microwave drying

Flower	Heating Time
Carnation	2 .5 to 3 minutes
Daffodil	1.5 to 2 minutes
Dahlia (3")	2 minutes
Miniature rose (three flowers)	70 seconds
Pansy - 2nd drying	1.5 minutes
Pansy - 1st drying	45 seconds
Rose	1.5 minutes
Sunflower	1.75 minutes
Verbena (four flowers)	75 seconds
Violet	1.5 to 2 minutes
Zinnia	2 to 2.5 minutes

Summary

Microwave drying is quick and relatively simple. It takes only a few minutes and provides dried flowers that look fresher and more colorful than obtained by other methods. Flowers with thick petals like magnolia are not suitable for drying in microwave. Flowers vary in moisture content, texture and density. Care should be taken to use the same sized flowers from one species at a time. Brightly colored flowers are best for drying. Flowers such as lilies, roses, violets, zinnias, and dahlias work well with this process. Drying multiple flowers does take longer, which can degrade their color. Flowers like marigolds, chrysanthemums, cornflowers and zinnias dry well in a fan-assisted, convection oven. Silica gel can be used in the microwave with fantastic results. The container is placed in the oven with a cup of water in the rear left corner of the microwave and the flowers are rearranged.

Questions and Answer

1. Give the procedure for drying in microwave oven.
2. Mention plants suitable for microwave drying.
3. What are the precautions to be kept in mind for drying in microwave oven?

Chapter-9

Freeze Dried Flowers

The process for freeze dried flowers began in 1989. Freeze drying was first used by the Peruvian Incas of the Andes, who stored their food on the heights of the Andes. The ancient Incas were the first to learn the benefits of sublimation. They discovered that when they dried food at high elevations, the results were better. Even though the food would freeze, it would dry well if left exposed to the sun. The reduced pressure at high elevation, combined with the dry air and energy from the sun, had caused sublimation to occur.

During World War II, freeze drying was commercially used to preserve blood plasma. Since then freeze dried products are used commercially and also for space programs. The preservation of flowers by this method was evolved in the 1990's. This is primarily because the flowers needed treatments before and after the process to avoid shattering and loss of colour (Tom Harris, 1999). By removing all the moisture from the flowers, the colours are preserved. As long as they are protected from exposure to sunlight, insects and humidity, they should last for years and years. Freeze dried flowers preserved from special occasions, evoke the past, and leave us with sentimental thoughts. They help us commemorate good times and offer comfort in sadder ones, punctuating every celebration in our lives. From weddings, anniversaries, births, the passing of a loved one, retirement, to Mother's Day and Valentine's Day, freeze dried flowers give us life long memories of the event, and continue to enrich our lives. When we savor these moments for ourselves, friends, and family, it only adds a higher quality of living for everyone.

Freeze drying (technically known as "lyophilization") is the process of lowering the temperature of an object and then using a vacuum to extract all the moisture from the item. It is a state-of-the-art technique and the most effective method for flower preservation today. It is an innovative vacuum process that takes approximately four weeks depending on the flower. Freeze dried flowers go through a process in which water is removed as vapor directly from ice, without passing through the liquid state. This process is called sublimation, and requires reduced pressure to occur. All other drying methods use evaporation. In other words, water is removed as vapor from liquid water with heated air. The freeze drying machine drops the temperature to a -20 degrees F then it slowly returns the freeze dried flowers to room temperature over a four week period. This slow preservation process allows the freeze dried flowers to retain their original form, while the colors become enriched.

Freeze drying is well suited for preserving sensitive biological materials. The three reasons for this is that, first, freezing slows or stops most chemical reactions. Second, the process occurs under a vacuum, and the absence of oxygen prevents oxidative reactions. And third, the process can be performed at very low temperatures, lower than any other drying method. When preserving sensitive biological materials, the main enemies are water, oxygen, and heat. Water is a powerful solvent, and is required for many chemical reactions, including enzymatic breakdown. Oxygen causes adverse chemical changes through oxidation, which is why anti-oxidant compounds are so popular. Heat can decompose or breakdown ingredients by itself, while at the same time accelerating other chemical changes, including enzymatic breakdown and oxidation. Each freeze dried flower is dried separately to ensure an aesthetically pleasing shape.

9.1 Method

The individual flowers are labelled and placed in the freeze-dry machine carefully, to avoid any bruises to the flowers. The stems are removed to enable the bloom to stand upright. An average machine holds approximately 2,000 heads of freeze dried flowers. After the four week period, the freeze dried flowers are delicately removed from the machine. This task usually takes about 4-5 hours. Each freeze dried flower is then dipped individually in a solution that seals the flowers and allows them to be soft and pliable. The freeze dried flowers are allowed to dry for a minimum of 36-48 hours. This additional process helps the freeze dried flowers to last indefinitely. The freeze dried flowers should be kept out of the direct sunlight and humidity to ensure their lasting quality. The hues of the freeze dried flowers may change slightly with time depending on the type and variety of the flower. The process for freeze dried flowers is a very labor intensive and time consuming process. However, this innovative technique allows your special flowers to last indefinitely.

9.2 List of Flowers for Freeze Drying

Flowers chosen for freeze drying should be of good quality with perfect shape and should have been picked freshly from the garden. Some important flowers for freeze drying are Anthurium, Gardenia, Ginger, Heliconia, Hyacinth, Liatris, Calla Lilly, Narcissus, Orchids, Tuberose and Bird-of-Paradise.

Summary

Freeze drying (technically known as "lyophilization") is the process of lowering the temperature of an object and then using a vacuum to extract all the moisture from the item. The process for freeze dried flowers is the state-of-the-art and most effective method for flower preservation today. It is an innovative vacuum process that takes approximately four weeks depending on the flower. Freeze dried flowers go through a process in which water is removed as vapor directly from ice, without passing through the liquid state. This process is called sublimation, and requires reduced pressure to occur. All other drying methods use evaporation in other words; water is removed as vapor from liquid water with heated air. Each freeze dried flower is dried separately to ensure an aesthetically pleasing shape. The freeze dried flowers should be kept out of the direct sunlight and humidity to ensure their lasting quality. The hues of the freeze dried flowers may change slightly with time depending on the type and variety of the flower. The process for freeze dried flowers is a very labor intensive and time consuming process. However, this innovative technique allows your special flowers to last indefinitely. The flowers suitable for freeze drying are Anthurium, Gardenia, Ginger, Heliconia,

Hyacinth, Liatris, Calla Lilly, Narcissus, Cattleya and Japhet Orchid, Tuberose, Bird-of-Paradise.

Question and Answer

1. What is Lyophilization?
2. What is the principle behind freeze drying?
3. Mention the process involved in freeze drying.
4. Mention plants suitable for freeze drying.

Reference

1. Tom Harris. 1999. How Freeze- Drying works. Website: http://recipes.howstuffworks.com/freeze-drying.htm

Chapter-10

Bleaching Flowers and Foliage

Bleaching is essentially whitening or decolourizing. It is an important step in the processing of dry plants. It involves chemical processes that can change the ability of colour bodies to absorb light by changing their degree of saturation (Robson, 1967). This chemical action removes the discolouration occurring during the dehydration phase of preservation of plants. It allows the effective use of dyes for colouring. Bleaching agent is a material that lightens or whitens a substrate through chemical action. To reduce the processing costs, multistage bleaching is suggested. In this method, both oxidative and reductive bleaching chemicals are used.

10.1 Sodium Hypochlorite Bleaching

The hypochlorites are one of the strongest classes of bleaching agents known. Sodium hypochlorite was used for bleaching at the inception of the commercial laundry industry in 1900. Earlier it was assumed that hypochlorites undergo bleaching by chlorhydrination of the double bonds in the plant material. Later, in 1980 it was reported that organo chlorine formation occurs during the hypochlorite bleaching process. The optimum condition for hypochlorite bleaching was at a pH of 10.5, accompanied with low temperature and a low hypochlorite concentration. The spent bleach liquor is an environmental pollution hazard unless disposed off correctly.

10.2 Hydrogen Peroxide Bleaching

Although, hydrogen peroxide was prepared as early as in 1818 by Thernard, it was used as bleach only in 1930. Roymoulik (1972) confirmed that prolonged exposure of cellulose to peroxide causes cellulose chain shortening or peeling due to loss of polysaccharide terminal sugars. Sjostrom (1981) quoted that bleaching to a high white, as required for ornamental plant material, needs extensive modification of lignin. The reactions between lignin and the chosen bleaching agent are thus important in commercial bleaching process. Therefore, the accessibility of the bleach to the lignin is the most vital factor to determine the bleaching efficiency.

Sjostrom (1981) also reported that the oxidative bleaches were more effective for bleaching plant tissues. The cleavage of double bonds breaks up the coloured molecules. This improves the solubility of molecular fragments. The chemical diffuses out of the tissue and into the bleach liquor. Hydrogen peroxide is inexpensive and readily available. It is commonly used in textile bleaching since it causes lesser damage to cellulose when compared to hypochlorites. 50 per cent of lignin can be removed by this bleach without much damage to cellulose.

10.3 Sodium Chlorite Bleaching

Masschelein (1979) affirmed sodium chlorite to be the most commonly used bleach for plant foliage. This is due to its selective mode of action. This bleach removed the entire colour from cellulose based material with minimal damage, even during prolonged contact. In the 1950s, sodium chlorite was used for bleaching after being introduced as a relatively stable solid surface of chlorine dioxide (Mark *et al.,* 1979).

Summary

Bleaching is essentially whitening or decolourizing. It is an important step in the processing of dry plants. It allows the effective use of dyes for colouring. Bleaching agent is a material that lightens or whitens a substrate through chemical action. The hypochlorites are one of the strongest classes of bleaching agents known. Oxidative bleaches like hydrogen peroxide are more effective for bleaching plant tissues. Sodium chlorite is also one of the commonly used bleach especially for plant foliage. Mostly bleaching agents are used after diluting with water in different ratios.

Question and Answers

1. What is the importance of bleaching in dry flower technology?
2. What are the different bleaching agents?

References

1. Mark,H. F., D. F. Othmer, C. G. Overberger and T. Seaborg. 1979. Dyes. In: Encyclopedia of chemical technology. Ed. A. Standen. Vol.3. John Wiley and Sons, USA: 151-211, 351-392
2. Masschelein,W. J. 1979. Chlorine dioxide: Chemistry and environmental impact of oxy chlorine compounds, Ann. Arbor. Sciencc, Michigan, USA: 89.
3. Robson, H. L. 1967. Bleaching Agents. Encyclopedia of Chemical Technology. 2nd Ed. Vol. 3. Ed. A.S tanden. John Wiley and Sons, USA. pp. 550-567.
4. Roymoulik, S. K. 1972. U.S.patent 3, 695,995, Oct 3.
5. Sjostrom,E. 1981. Wood chemistry, fundamentals and applications. Academic Press, New York, USA.

Chapter-11

Natural and Chemical Dyes

The major drawback of dried arrangements and other dry flower products is that they are dull and lack the brightness of fresh flowers. To overcome this problem, flowers are bleached and dyed. This enhances the visual appeal of the product. Both chemical and natural dyes can be used. Squires (1971) reported on the use of various types of paints to colour dried foliage and flowers using brushes or by dipping and spraying. Enamel paints, interior paints, poster paints and tube paints could be employed for this purpose. Tampion and Reynolds (1971) explained the three ways by which dyeing can be carried out for fresh flowers. They are as follows:

- By absorption i.e. when the cut stems are placed in a dye solution.
- By dusting the cut blooms into a solution of dye to which a few drops of wetting agent is added. This increases the uptake of the dye. They recommended dyeing of carnation, narcissus and chrysanthemum by this method.
- By using culinary dyes, which are often used to colour icing sugar. They are available in a wide range of colours and non toxic.

11.1 Chemical Dyes

Basic dyes were the first of the synthetic type to be made out of coal tar derivatives. Although basic dyes produce brilliant colours, they have poor fastness to light and washing. The direct dyes do not produce very brilliant colours but have better fastness to light and washing. Colour fastness of acid dyes is generally increased by adding metallic salts to the dye stuff. The sulphur dyes have excellent resistance to washing but not to sunlight. Azoic dyes or the naphthols are primarily used for bright red shades. The vat dyes are known for their fastness to both washing and sunlight. They are made from indigo anthroquinone and carbazol. The reactive dyes can be used such that the whole spectrum of colour can be put onto cloth using just one class of dyes. The other textile dyes include acetate dyes, alizaric dyes, aniline block, chrome dyes, soluble vats, neutral dyes, disperse dyes and indigo. Acid dyes are water soluble anionic dyes, while the vat dyes are water insoluble dyes. The basic dyes and diazo direct dyes are acutely toxic. The flowers of gypsophila and hydrangea can be made to absorb the dye from the dye solution mixed with water. Dip dyeing and spraying is recommended to colour the seeds and pods. Natural dyes and commercial dyes may also be used.

11.2 Natural Dyes

An upsurge in interest in natural dyes has been manifested as reconstruction of ancient and traditional dyeing technology, growing use of natural dyes in craft work and replacement of synthetic dyes for food safety. Natural dyes have become a part of human life since time immemorial. These dyes obtained from natural sources are non-pollutant, non - allergic, having rich shades (Gupta et al., 1997). Due to carcinogenic nature of synthetic dyes and their intermediaries there is enough scope to revive the age old art of natural dyeing.

11.3 Sources of Natural Dyes

India is a store house for large number of indigenous flora, which is a good source of natural dyes. The natural dyes cover a wide range of chemical classes viz., alkaloid, anthroquionone, aurone, benzoquionone, carotenoid, chalcone, chlorophyll, diaryol methane, flavone, flavonol, flavonone, indigoid, napthoquionone and quionone. Some sources of the dye colour are mentioned below (Gupta *et al.*, 1997).

Plants yielding Blue dyes

S.N.	Botanical Name	Common Name	Family	Plant part
1.	*Acacia nilotica*	Sunt berry	Leguminoseae	Seed pods
2.	*Cassia auriculata*	Tarwar	Caesalpinaceae	Bark
3.	*Mangifera indica*	Mango	Anacardiaceae	Bark
4.	*Malphigia glabra*	Barbados cherry	Malpighiaceae	Bark
5.	*Nymphaea alba*	Water lily	Nymphaeaceae	Rhizomes
Plants yielding Red Dyes				
1.	*Beta vulgaris*	Beetroot	Chenopodiaceae	Root
2.	*Caesalpinia sappan*	Patang	Caesalpinaceae	Wood chips
3.	*Erythrina variegate*	Coral tree	Leguminaceae	Bark
4.	*Plumbago indica*	Chitrak	Plumbaginaceae	Roots
5.	*Punica granatum*	Pomegranate	Punicaceae	Flowers
Plants yielding Yellow Dyes				
1.	*Curcuma longa*	Turmeric	Zingiberaceae	Rhizomes
2.	*Lawsonia inermis*	Henna	Lythraceae	Leaves
3.	*Nyctanthes arbor-tristis*	Harsingar	Nyctanthaceae	Flowers
4.	*Tectona grandis*	Teak	Verbenaceae	Leaves
5.	*Terminalia chebula*	Harda	Combretaceae	Fruits

According to Mahanta and Tiwari (2005), Arunachal Pradesh, is one of the hotspots of biodiversity and is home to a range of economically important plants. Some of these plant species have found use in the preparation of natural dyes. There are more than 500 dye-yielding plant species in this state. Natural dyes are environment friendly; for example, turmeric, the brightest of naturally occurring yellow dyes is a powerful antiseptic and revitalizes the skin, while indigo yields a cooling sensation. Research has shown that synthetic dyes are suspected to release harmful chemicals that are allergic, carcinogenic and detrimental to human health. Thirty-seven species belonging to 26 families have been recorded in the present work. Fabaceae is found to be dominant with six species followed by Euphorbiaceae and Moraceae with four each and Juglandaceae with two species. The remaining 21 families have one species each. Various plant parts, viz., leaves (*Aporusa diocia*), bark (*Bischofia javanica*), flowers (*Clitoria mariana*), fruits (*Juglans regia*), seeds (*Bixa orellana*), etc. of the recorded species were found to be employed traditionally by the tribal communities for extracting dyes utilizing indigenous extraction techniques.

Possible sources of dyes are:

Source	Part used	Source	Part used
Apple	bark, twigs	carrot	Peels, tops
asparagus	mature plant, seed pods	cucumber	peels
beets	peels, tops	deadly nightshade	berry stem, leaf, root
blue grass	seed heads	geranium	Flowers

Purohit *et al.*, (2007) developed multiple natural dyes from flower parts of gulmohar. The different shades achieved with dyes extracted from different parts of the plant are described below:

Parts of the flower	Time of extraction	Solubility	Mordants	Shades
Whole flower	3 hours	0.5 % NaOH	Turmeric powder	Golden yellow
Whole flower	6 hours	0.5 % NaOH	10per cent Alum	Olive green
Petal (part I)	3 hours	0.5 % NaOH	Turmeric powder	Dark tan

Parts of the flower	Time of extraction	Solubility	Mordants	Shades
Petal (part I)	3 hours	0.5 % NaOH	10 % Alum	Olive green
Petal (part I)	3 hours	0.5 % NaOH	Nil	Saddle brown
Petal (part II)	6 hours	0.5 % NaOH	10 % Alum	Olive green
Petal + reproductive organ (part I)	3 hours	0.5 % NaOH	Nil	Dark brown
Petal + reproductive organ (part II)	6 hours	0.5 % NaOH	Turmeric powder	Brown
Calyx	---	0.2%	Nil	Brown

11.4 Natural Dye Extraction Methods

Extraction methods for natural dyes include drying, pounding, soaking, skimming crystallizing, condensing, caking, liquefying etc. depending on the quality and species of the dye yielding plant (Shrivatsava and Dedhia, 1999).

Traditional fermentation process: This method is used for extraction of indigo, annatto etc. The process involves the disruption of plant material followed by acidified aqueous extraction for chlorophyll, carotenoids and annatto. Overnight fermentation of indigo plants using bacteria is done in specially designed cement moulded tanks. The water should be free from lime salts, which cause precipitation of indigo. The slurry settled on the bottom is collected, boiled and sun dried.

Enzyme aided extraction: Enzymes are used to facilitate the extraction, increase pigment yield and improve the stability of the resultant colour preparation. E.g. when orange peels were pretreated with crude microbial enzyme preparation, the carotenoid pigment increased by 46 to 65 percent (Elias *et al.*, 1984).

Microorganisms aided extraction: Natural dye extracts, especially those obtained through aqueous extraction or direct pressing of plant materials, contain pigments along with a high proportion of unwanted dissolved solids like salts, organic acids, phenols and sugars. Approximately, 80 percent of beet juice solids are carbohydrates and nitrogenous compounds. Fermentation of juice using yeast (Saccharomyces sp.) and mould (Aspergillus sp.) reduced the solids substantially. About 5 to 7 fold increase in pigment content on dry weight basis was observed (Avila *et al.*, 1983).

Supercritical carbon di oxide extraction: The use of super critical carbon di oxide as an extracting solvent is one of the modern approaches to avoid the detrimental effect of heat and light and also to reduce the interference of co-extracted material. E.g. The pigment bixin has been extracted from annatto using this method (Chao *et al.*, 1991).

Indigenous extraction methods: Fresh, mature fruits and barks of *Terminalia chebula* and *Emblica officinalis* are used for dye extraction. These parts are taken in almost equal parts by weight, crushed and boiled in water for long periods in iron pots. The boiled mixture is then transferred to a pot having minute pores at the bottom through which only the thin liquid can percolate, leaving behind the residues in the pot. The pot with the mixture is placed over a tripod stand and another pot is placed below it. The black thick liquid from the upper pot is collected in drops and allowed to cool down. The thick liquid so collected in the lower pot is the dye (black) used for colouring clothes Mahanta and Tiwari (2005).

In another process, the whole plant and leaves of *Rubia cordifolia* or *R. sikkimensis* along with shoots of *Ficus altissima* separately or with parts of other natural dye-yielding plant species like *Juglans regia, Woodfordia fruticosa, Illicium griffithii*, etc. are crushed and put in an earthen pot. To this little water is added. The pot is kept undisturbed for 20-25 days during which period the contents of the pot get fermented. The fermented content is boiled to get a thick liquid and the extract is filtered through a piece of thin cloth to yield the natural dye. The colour of the dyes thus extracted and prepared through this process depends upon the plant species and parts used.

11.5 Mordants

Natural dyes fade easily because they do not fix well. Mordants are substances which are used to fix dyes to the material. They improve the take up quality and help improve colour and light fastness. Traditionally, mordants are found in nature (Anon., 2007). Some of the chemical mordants being used now are:

- Iron (Ferrous sulphate)
- Tin (Stannous chloride)
- Chrome (Potassium dichromate)
- Copper sulphate
- Tannic acid
- Oxalic acid

Some natural mordants are juice of unripe grapes, vinegar, clay, lime, wood ash, animal urine etc. *Phyllanthus emblica, Terminalia chebula, Juglans regia, Terminalia arjuna, Acacia nilotica,* and *Eucalyptus sp.* are also natural mordants.

The following methods are practiced for applying mordants :

- Pre application of mordants: The material to be dried is kept in the mordant for 24 hours. Then dyeing is done.

- Mordanting while dyeing: Mordants are mixed along with dried material and dye simultaneously.
- Post application of mordants: The material is first dyed and then immersed in the mordant solution.

Summary

The major drawback of dried arrangements and other dry flower products is that they are dull and lack the brightness of fresh flowers. To overcome this problem, flowers are bleached and dyed. This enhances the visual appeal of the product. Both chemical and natural dyes can be used. There are three ways by which dyeing can be carried out for fresh flowers. They can be done by absorption i.e. when the cut stems are placed in a dye solution, by dusting the cut blooms into a solution of dye to which a few drops of wetting agent is added and by using culinary dyes which are often used to colour icing sugar, owing to their wide range of available colours and non toxicity to plants. Mordants are substances which are used to fix dyes to the material. They improve the take up quality and help improve colour and light fastness. Some natural mordants are juice of unripe grapes, vinegar, clay, lime, wood ash, animal urine etc. *Phyllanthus emblica, Terminalia chebula, Juglans regia, Terminalia arjuna, Acacia nilotica,* and *Eucalyptus sp.* are also natural mordants. Mordants can be applied by pre application of mordants, mordanting while dyeing and post application of mordants.

Question and Answer

1. Name a few plants yielding natural dyes.
2. What are the different natural dye extraction methods? Explain.
3. What is a mordant?
4. What are the methods of applying mordant?

References

1. Anon. 2007. Natural dyeing of textiles. Website: http:/ / www.itdg.crg
2. Avila, A., F.Garcia- Henandez and E. Santos. 1983. Fermentation of an aqueous extract of Mexican beets *Beta vulgaris* to obtain red colourants. Developments in Industrial Microbiol., 24: 553-561.
3. Chao, R.R., S.J. Mulvaney, D.R. Sanson, Hsieh, Fu and M.S. Tempasta, 1991. Supercritical CO_2 extraction of annatto (*Bixa orellano*) pigments and some characteristics of the colour extracts. J.Food Sci., 56:80.

4. Elias, A.N., Foda, M.S. and Altia, L. 1984. Production of pectin and pigments from orange peels by using microbial enzymes. Egyptian J. Food Sci., 53: 1582.

5. Gupta,V.K., R.A. Sachan, V.P. Singh, and S.K. Sharma, 1997. Natural Dyes. Indian Textile Journal, 108:16.

6. Mahanta, Debajit and S.C. Tiwari. 2005. Natural dye yielding plants and indigenous knowledge on dye preparation in Arunachal Pradesh, Northeast India, Current Sci., 88 (9): 1474-1480.

7. Purohit, A., S. Mallick, A. Nayak, N.B. Das, B. Nanda and S. Sahoo. 2007. Developing multiple natural dyes from flower parts of Gulmohur. Current Science, 92 (12):1681-1682.

8. Shrivatsava, A. and Dedhia, E.M.1999. Extraction and dyeing methods for natural dyes. In: Natural dyes, Scope and Challenges. Eds. M.Daniel, S.D. Bhattacharjee, Arun Arya and Vinay, M. Raole 2006. Scientific Publishers (India), Jodhpur. pp 7-34.

9. Squires, M. 1971. The art of drying plants and flowers. Gramercy publishing Company, Crown Publishers Inc., USA. pp.245.

10. Tampion, J. and Reynolds, J. 1971. Botany for flower arrangers. Pelham Books Limited, London, pp. 175.

Chapter-12

Techniques to Dye Dried Flowers and Foliage

One of the important aspects of dry flowers is the pleasure of gathering flowers and grasses to make wonderful dried arrangements. The only drawback is that the original colors fade and the floral arrangements can end up looking like a bunch of straw. To enhance the colors of dried flowers, dyeing is advocated. While coloring ferns or small leaves, it is better to work on an aluminum foil or waxed paper so it can collect extra paint, which can be used on another leaf. Leaves should be lightly brushed, and never saturated with paint. The painted specimens are placed on the waxed paper or on a computer paper to dry. Newsprint should not be used.

Dyed or coloured material always lasts longer than those that are not touched up with color. Keeping the coloured plant materials away from direct sunlight helps retain the color up to six to eight years. Very thin flowers should not be coloured because they might roll up and disintegrate. Many flowers actually do not need dyeing because they have naturally lasting colors. Flowers like Delphinium and Larkspur can hold their colors for years if kept away from sunlight. Blue Scilla also holds its blue color. Forget-me-nots are occasionally pink when they are dried but they retain their soft blue as well. Christmas Cactus, Cyclamen, Marigold, Buttercups and Coreopsis retain their colors as well. However, most wild flowers are too thin to be successfully color washed.

Basically dried flowers need to be out of the contact with direct sunlight while they are placed for decorative purposes. This is primarily because direct sunlight speeds up the fading process of these beautiful dried flowers to a large extent. Similarly, dried flowers should not be placed in the path of the forced air from a cooler or an air-conditioner or some other heat source. The air gushing out from such appliances are awfully dry which initiate a shattering process in the petals of the dried flowers. The most preferable forms of dried flowers you may use are wreaths, both vertical and horizontal swags, bouquets, sheaths and sprays. Another very appealing way of using dried flowers for decorative purposes is by just hanging them around the room. Some in bunches in a vase or some in clusters in a basket is also appealing and can be quite innovative and attractive. These dried flowers need to be dusted on a daily basis using a soft, feather duster or a hair dryer setting it on the lowest. It is always good to wrap the flowers in a newspaper or placing them in a cardboard box. The box should not be placed anywhere near damp or even overtly dry places.

12.1 Dyeing Techniques

Dyes are organic compounds, which absorb light of certain frequencies, thus only reflecting a portion of the visible spectrum with the result that the eye sees color. Each pure dye compound therefore has a unique color. The vast number of different colors available is the result of blending pure dyestuffs. There are several classes of dyes, which can stain plant material. Food dyes are widely used for this purpose. Although relatively safe to handle, food dyes are rarely lightfast and are only available in a limited range of color. They are more expensive than other dyes because of the rigorous and costly toxicity testing required for food additives. There are different types of dyes that are used to dye dried florals and make them more beautiful.

- A dye bath, rubber gloves, glass jug, a stirring utensil, and paper towels are required.
- Paper gloves should be used before handling dyes.
- Dyes should be dissolved in hot water.
- The dye bath is filled with very hot water and dye dissolved in it.
- The quantity of dye needed depends on the depth of color required.

- Dried flowers or grasses are completely wetted and submerged in dye bath.
- The dried flowers are removed from the solution once the desired color is achieved.
- The dried flowers are rinsed carefully and allowed to dry by laying them on paper towels.
- During drying process, the dried flowers or grasses are tapped gently to separate.

12.2 Dyeing Tips

Dyeing is done in hot water dye bath because the surface of most plant fibers is covered with a thin layer of wax called the cuticle. Penetrating this layer is difficult at cold or warm temperatures. Heat breaks down the wax surface and also swells plant fibers allowing for uniform dye migration within the substrate.

Dyes are sold in concentrated powder form. The color of dyed flower depends upon the base color of the material as well as the amount of dye taken up by the fiber. This is termed as depth of shade. Depth of shade is usually expressed as the amount of dye applied to goods computed as a percentage on weight of the goods.

Summary

To enhance the colors of dried flowers, dyeing is advocated. While coloring ferns or small leaves, it is better to work on an aluminum foil or waxed paper so it can collect extra paint, which can be used on another leaf. Dyed or coloured material always lasts longer. Very thin flowers should not be coloured because they might roll up and disintegrate. Basically dried flowers need to be out of the contact with direct sunlight while they are placed for decorative purposes. Dyeing is done in hot water dye bath.

Question and Answer

1. Explain the techniques to be followed while dyeing dry flowers.

Chapter-13

Methods to Create Skeleton Leaves

Skeleton Leaves

Skeleton leaves are semi-transparent leaves, also called fossil leaves. They are prepared by soaking the fresh leaves in bleached water and then by hand rubbing the soft green flesh from the network of veins. The leaves can be then dyed to variant colors to give them a beautiful and pleasing appearance. Although delicate by nature, they are actually sturdy and can easily be glued to a variety of surfaces. They come in several colors and sizes to meet a complete line of crafting needs. The delicate beauty of these leaves is sometimes meant to capture life, hope and memories. These delicate, gorgeous skeleton leaves are a perfect accent for all craft projects. They make a beautiful and elegant statement when they adorn your wedding invitations. They work surprisingly well on gift tags, greeting cards, scrapbooks, collages, papermaking, stenciling, and stamping. They are perfect for decorating bridal bouquet, wedding reception table, candles, and wedding favors.

Skeleton leaves can also be made by placing selected leaves in a pot with equal amount of bicarbonate and baking powder. Water is added in the pot, boiled and stirred gently on a low flame. Once leaves have softened, the chlorophyll is rubbed off gently. Then the leaves are bleached and dried on paper towels. These skeleton leaves can be dyed with a colors to give them a pleasing and an attractive look. The skeleton peepul leaf is mounted on various objects like handmade paper, scrapbooks, wedding cards, etc. to give it an elegant and authentic look.

Magnolia Leaves Skeleton

The blade of a magnolia leaf is strengthened by a network of veins, which may be parallel or may branch out from the central midrib.

Each vein contains vascular bundles, which are a system of tubes for carrying water and food to the leaf cells. When a leaf falls from the tree the cells get broken down by insects feeding on them and by bacterial decay. The tougher veins persist as the leaf skeleton. Magnolia leaves can also be subjected to a chemical process to remove all the juice and sap and leave only skeleton veins. They are then colored according to the seasons and overprinted with a slogan or message. Magnolia leave skeleton is a beautiful decorative piece and is used for decorative and ornamental purposes.

Mexicana Leaves

Botanical Name : *Ulmus mexicana*

Common Names : Tirrá, palo baqueta, Mexican elm

Mexicana tree is one of the spectacular forest trees, having a clear straight bole, unbuttressed and generally unbranched until near the top. The leaves are simple and alternate with an elliptical or ovate shape. The edges of the leaves are serrated and they bear petioles. The leaves are glossy, having prominent midribs and veins in the lower surface. Dried Mexicana leaves are beautiful for crafting projects. They are used to decorate books, scrapbooks and other similar stationery. They can be colored with vibrant colors and used for embellishing frames, packing, etc.

Peepul Leaves

Botanical Name : *Ficus religiosa*

Common Names : Bodhi, Bio Tree, Pipul, Sacred Tree

Peepul tree is the most well known tree worldwide. It represents three major Gods- Bhrama, Vishnu and Shiva. To plant a peepul tree is considered to be a sacred deed. It is a medium sized tree with a relatively short trunk and a large crown with wonderful wide spreading branches.

The leaves of peepul tree have a beautiful heart shape and they taper to a needle point. When the leaves first appear their color is red-pinkish but then turn deep green and grow to about 12 to 18 cm long. The foliage can often be dense.

Peepul Leaves Skeleton - The leaves of the sacred peepul tree have beautiful heart shape and they make a rustling sound when they brush against each other. When the leaves dry naturally, they leave behind a webbed skeleton of veins.

Summary

Skeleton leaves are semi-transparent leaves, also called fossil leaves. They are prepared by soaking the fresh leaves in bleached water and then by hand rubbing the soft green flesh from the network of veins. The leaves can be then dyed to variant colors to give them a beautiful and pleasing appearance. Some of the leaves useful for skeleton leaves blade of a magnolia leaf, dried mexicana leaves, peepul leaves skeleton etc.

Question and Answer

1. What is skeletonising of leaves?
2. How will you skeletonise peepul leaves?

Chapter-14

Annual Plants for Dry Flower Making

Flowers blooming in a garden can be preserved by drying them for use in arrangements, decorations and various other applications. Natural dry flowers can be annual or perennial. Some of the important annuals used for dry flower making are described below.

14.1 Annual Flowers

Plants classified as annuals are those that complete their life cycle within the space of one year. Annual flowers are versatile and sturdy. The annuals have many uses in landscaping. Annual flowers are beautiful and garden lovers tend to preserve these flowers throughout. They dry these flowers and store them for use in several beautiful dry flower arrangements.

List of Annual flowers for dry flower production

Acroclinium	Land Lotus Flower
Amaranthus	Nigella
Winged Everlasting	Marigold
Anemones	Matricaria
Armerias	Millet Flowers
Bachelor's Button	Orchid Flowers

Blue Salvia
Broom Bloom
Celosia
Chamomile
Dahlia Flowers
Gomphrena
Feverfew
Helichrysum
Helipterum
Hydrangea Flowers
Oregano
Chinese Lanterns
Phalaris
Purple Majesty
Rudbeckia
Sea Holly
Strawflower
Sweet Annie
Verbena
Zinnias

Amaranthus

Botanical Name	:	*Amaranthus* sp.
Common Names	:	Tampla, Cat Tail, Love Lies Bleeding
Colors	:	Red, Green, Cream
Cultivation	:	Full Sun
Growth habits	:	Annual, the size depends on the cultivar, upto 5 feet tall
Uses	:	Used in accenting dried flower arrangements
Regions	:	Warmer South America, Mexico, Dakota

Amaranthus is a beautiful flower used in several dried flower arrangements. There are two varieties of Amaramthus that can be dried and preserved for longer periods: *Amaranthus Caudatus* and *Amaranthus Hypochondriacus*.

Drying Amaranthus

Drying of *Amaranthus Caudatus* – This plant has large, soft stems with broad leaves, producing trailing red flowering stalks. Trailing dark red spikes with good color, on long stems are ideal for drying. Older blooms will fade. Spikes with three or four large stems per bundle having multiple heads should be hung to dry. Mature spikes can be saved for seed. Spikes of this plant are often treated with glycerin and burgundy red dye to give a long lasting and durable novelty flower.

Drying of *Amaranthus Hypochondriacus* -This plant is characterized by upright red or green flower spikes on soft, leafy stems. Stems should be picked early to avoid shattering of over mature stalks and hung dry in bundles of 5 - 7 stems. Mature spikes can be saved for seed.

Cornflower

Botanical Name	:	*Centaurea cyanus*
Common Names	:	Cornflower, Dwarf Blend, Blueboy, Black Magic, Blueball, Bachelor's button, Bluebottle, Boutonniere flower, Cyani flower Hurtsickle.
Cultivation	:	Full sun with almost all types of soils
Bloom Time	:	Summer, Autumn
Colors	:	Pink, White, Light Blue, Blue, Violet
Uses	:	Elegant in vase and other decoration purpose
Regions	:	Cold Nothern Regions of India, Africa, North America

Cultivating cornflower -Cornflower is a very easily grown plant and succeeds in an ordinary garden soil. They can even be grown in walls. These flowers are usually self sowing. The leaves have a refreshing aromatic fragrance. The stem is upright, erect, hairy, finely furrowed and branching. Leaves are alternate, hairless, toothed, and divided into broad, lobed segments.

Drying cornflower : They have low moisture content and so they air dry very well. The flowers are picked when they are just about to bloom and bunched together. Each bunch at the bottom is bound with string or rubber bands. Bunches are hung in a relatively dark, airy spot. A fan can be used to increase air circulation.

Uses -They are one of the easiest flowers to dry for everlasting arrangements and for beautiful winter decoration. The dried petals are used in potpourri in order to add color. The dried flowers are antipyretic, astringent, antitussive, weakly diuretic, ophthalmic, emmenagogue, very mild purgative and tonic. An infusion can be used in the treatment of dropsy, constipation, or as a mouthwash for ulcers and bleeding gums. The dried flowers and petals are used as a flavoring in cooking to give food a deliciously aromatic bitter taste. The dried flower buds are used as an insecticide.

Blue Salvia

Botanical Name	:	*Salvia farinacea*
Common Names	:	Victoria Blue Salvia
Cultivation	:	Full sun in well drained soil
Colors	:	Blue, Purple, Red, Pink, Lavender
Bloom Time	:	December to May
Uses	:	It is one of the few true blue dried flower accents
Regions	:	Arid Regions of India, Spain, North Africa, North America

Drying Salvia : These are beautiful flowers that are preserved for enhancing the decor of homes and offices. Spikes are picked when the florets are fully open and the calyx is bright blue. Then these stems are tied together and hung upside down to dry in a dark, well ventilated place. After they are completely dried they are taken out and stored in an air tight container.

Uses - Dried Salvia are used as an everlasting in exclusive dried flower arrangements. They are paired with a variety of white porcelain and used to make beautiful and charming bouquets. Dried leaves within clothes and linen will discourage moths.

Broom Bloom

Botanical Name	:	*Cytisus Scoparius*
Common Names	:	Round Bloom, Spanish Bloom
Cultivation	:	Full sun in very well drained soil

Colors	:	Royal Blue, Yellow, Green, Burgundy
Bloom Time	:	July
Uses	:	An excellent dried flower filler offering an array of color
Regions	:	Northern Asia, England, Europe

Cultivation of Broom Bloom -Broom is most easily raised from seed. They are sown, broadcast in the open air as soon as they ripe. Seedlings are transplanted in autumn or spring to their permanent position. Their roots strike down deeply into the ground and so the plants are grown in dry, sandy soil. They do well on rough banks.

Drying Broom

Broom bloom is an excellent flower to dye. These are dyed before drying. Immersion dyes are used for dyeing broom blooms. These flowers are cut when flowers are completely opened, collected and made into small bunches. Each bunch is tied with a thread or rubber bands and hung upside down in dry, dark, open, airy place. When completely dried, they are removed and preserved carefully.

Uses : Dried brooms are a pre eminent flower for dried flower arrangements, in wreaths, bouquets and baskets. It is one of the favourite dried flower filler. Broom tops are dried and infused with boiling water for treatment of several diseases. The dried leaves are used as medicines. In early times these flowers were pickled and preserved in salts and eaten as capers.

Celosia

Common Names	:	Woolflower, Redfox
Botanical Name	:	*Celosia argenata*
Cultivation	:	Full sun with dry to moist soil
Colors	:	Red, Orange, Yellow, Pink, Magenta
Bloom Time	:	April to October
Uses	:	In arrangements, wreaths, and swags
Regions	:	Cool mountainous regions, Tropical Regions of North America, China, South East Asia

Cultivation of Celosia : Celosia thrives in well drained soil with a pH of 6.0 to 6.4. They grow well in locations with a rich, deep, loamy soil and at least 8 hours of full sunlight. Plants with too little sunlight will not produce good flowers. These flowers require 4 to 12 weeks of vegetative growth prior to flower initiation for optimum shoot development. They require relatively high amounts of nutrients, nitrogen is the key element. Fertility levels should be raised before planting celosia.

Drying Celosia : Celosias are well suited for drying and selling as everlasting cut flowers. Wheat type celosia has a tendency to shatter when dried, making it difficult to handle or ship without damaging the flower head. For the best color retention, celosia should be dried rapidly. Some use a greenhouse covered with black plastic for drying. Stems are harvested and are placed on a bench under the greenhouse. Exhaust fans are used to carry out moisture. They are dried for about 7 to 14 days.

Chinese Lanterns

Botanical Name	:	*Physalis alkekengi*
Common Names	:	Ground Cherry, Winter Cherry
Cultivation	:	Moderate sunlight with moderate soil
Colors	:	White
Bloom Time	:	Mid Summer
Uses	:	Attractive in vases for dried arrangements
Regions	:	South eastern India, China, Europe, North America

Drying Chinese Lanterns : The stems are placed on a horizontal position, allowing the lanterns to droop sideways. Stems should be cut when the pods become orange in late summer. Collect the stems and then remove leaves to dry them. After removing the leaves, hang them upside down in a shady place. They can also be dried by microwave drying.

Dahlia Flowers

Botanical Name	:	*Dahlia* spp
Common Names	:	Dahlia
Cultivation	:	Full sun with well drained humus soil
Origin	:	Native to Mexico and South America
Colors	:	Pink, Red, Orange, Purple, Cream, White, Scarlet, Magenta
Bloom Time	:	June to October
Uses	:	As wonderful dried flower accent
Regions	:	South America, India

Cultivation of Dahlia : Dahlia likes well-drained, humus rich soil. They also like full light. They can't grow under cooler climate with plenty of rainfall. Tubers are planted after dusting them with fungicide. After the growth reaches between a half inch to a full inch, the plant is again dug up and clump divided into a few pieces. Each new clump needs to have a shoot on it, to be used as a planting material.

Uses : Dried dahlia allow us to extend our garden pleasures. They are the most beautiful flowers used as everlastings in a dried arrangement. They can enhance the beauty and cheerfulness of any dried arrangements. They are best for all sorts of arrangements like wreaths and bouquets. They are also useful as beautiful centerpiece placed in a vase or decorative bottle. They are widely used for decorations and medicinal uses.

Gomphrena Flower

Botanical Name	:	*Gomphrena globosa*
Common Name	:	Globe Amaranth, Bachelor's button
Cultivation	:	Full sun with well drained soil
Colors	:	Purple, Pink, Lavender, Orange
Uses	:	Used widely in homemade potpourri

Gomphrena produces clover like flowers in cream, pink, red and raspberry colors. It likes average, well drained soil and is tolerant of drought, but does just fine in high humidity as well. Flowers will dry right on the plant. The flowers have to be harvested, once they feel light and papery. Then they are air dried for few days until they are fully dried. They can be bleached and dyed with various colors and used in several dried floral arrangements.

Uses of Gomphrena

Dried gomphrena has been widely used in making potpourri. To make a pot pourii dried leaves and flower petals are placed into a large bowl. Several spices and dried seeds are mixed in small amounts to attain fragrance. A fixative is added to it and shaken properly. The container is filled with dried gomphrena and secured using netting along with a ribbon.

Feverfew

Botanical Name	:	*Chrysanthemum parthenium*
Common Name	:	Feverfew
Cultivation	:	Full sun in average to poor soil
Colors	:	White, Off-White
Uses	:	Medicinal uses
Regions	:	Europe, Australia, Southeast America

Cultivation of Feverfew

Feverfew is a herbaceous plant. When once planted it gives year after year an abundant supply of blossoms with only the merest degree of attention. It is a very easily grown plant and succeeds in an ordinary garden soil. They can even be grown in walls. These flowers are usually self sowing. The leaves have a refreshing aromatic fragrance. The stem is upright, erect, hairy, finely furrowed and branching. Leaves are alternate, hairless, toothed, and divided into broad, lobed segments.

Uses : Dried feverfew are in great demand in floral industry. Dried leaves are also used in tea for a good flavor. It is also preserved and added to certain foods to give flavour to the foodstuff. Dried leaves of feverfew are used in the treatment of migraine.

Drying Feverfew

Feverfew should be dried appropriately for subsequent processing while preserving the active ingredients. They can be easily dried by the method of air drying. They should be cut just before the flowers are in full bloom and then hanged upside down in dry, dark, open, airy place. If required, a fan can also be used to enhance the process of drying. After drying, they are stored in a air tight container until further use.

Helichrysums

Botanical Name	:	*Helichrysum bracteatum*
Common Name	:	Everlasting
Meaning	:	Gold
Cultivation	:	Full sun in well-drained and slightly moist soil
Colors	:	Bright Yellow, Cream, Gold
Bloom Time	:	Mid Summer
Uses	:	Dried and exotic arrangements
Regions	:	Forest margins of India, South Africa

There are over 600 species of helichrysum occurring worldwide with 245 found in Southern Africa. The word Helichrysum is derived from the Greek "helios" meaning sun and "chrysos" meaning gold, referring to many of the flowers of species in this genus. The most commonly used species of helichrysum are *Helichrysum psilolepis, H.benguellense, H. umbraculigerum, H.nudifolium* and *H. platypterum*

Cultivation of Helichrysums

This is an easy to grow plant requiring very little maintenance. They are planted in full sun and well-drained soil. They can be raised from seed; various species may be propagated by cuttings, inserted in spring. To shape or encourage new growth, the plant can be pruned lightly after flowering.

Drying of Helichrysum

Flowers for drying should be collected when partially unfolded and suspended. These are collected and tied together. They should be then placed with their heads downwards in a cool place until they dry. After drying they are stored in air tight containers.

Uses of Helichrysum

Dried helichrysums are used mainly for dried and exotic arrangements. The colorful medley of these flowers is great for everlasting dried flowers. These dazzlers radiate colors all over the place giving a shine to the place. They also have food, ornamental and medicinal value. The dried leaves are used to make an ointment to be applied on the body.

Helipterum

Botanical Name	:	*Helipterum roseum*
Common Name	:	Pink Paper Daisy, Pink Everlasting
Cultivation	:	Partial shade to full sun in average soil
Colors	:	White, Pink
Bloom Time	:	September
Uses	:	Nice dried flower for a bright accent
Regions	:	Indian Ocean Islands, East India, Tropical and warm temperate regions, Australia, South Africa

Cultivation of Helipterum

Helipterum means "plumed pappus" and are best grown in partial shade to full sun in average soil. The seeds are sown directly in autumn as the soil cools down. Plants grow slowly thorough the winter months, developing a strong root system. The flowers later bloom to its full in spring.

Drying of Helipterum

Pink everlastings are picked up when the flowers are picked at very young stage when about to bloom. If the flowers are too late they tend to open too much and become unattractive. They are collected and bunches are made out of it. These bunches are tied together and hanged upside down in a dark and well-ventilated place. After drying, the flowers are untied and carefully stored in a safe place away from direct sunlight. These flowers are very fragile and so good care has to be taken while preserving and storing these flowers. They have to be handled with care.

Uses of Helipterum

Dried helipterum complements any home forming beautiful dried arrangements. They are used as wonderful cascading over walls, hanging in baskets or as decorative pieces in pots and vases. They are also used in permanent posy displays.

Hydrangea Flowers

Common Name	:	Snow Ball
Botanical Name	:	*Hydrangea macrophylla*
Cultivation	:	Partial shade to full sun in well-drained soil
Colors	:	Blue, Pink, White
Bloom Time	:	March, April, May
Uses	:	Offer a dramatic, large head dried flower
Regions	:	Coastal regions of India, Mountain regions of Japan, Korea

Cultivation of Hydrangea Flowers

Hydrangea flowers are easy to grow on well drained soil, which should contain plenty of organic matter or humus. They prefer morning sun and afternoon shade. Planting should be avoided in hot, dry, exposed sites. They respond to several light applications of fertilizer during the growing season. After the application of fertilizers, water should be applied to help dissolve the fertilizer and send it into the soil.

Drying of Hydrangea Flowers

Dried hydrangeas have become very popular in recent years. The stage of harvesting varies from crop to crop and hence cannot be specified. The flower heads dry well when gradual air drying is done. However the weather should not be too humid. These flowers are unique in that they will change their natural colors to shades of plum, mauve, steely blue, dusky rose, amber, brick red, sea green, and other antique colors according to the soil pH. To dry these flowers, the flower head is cut with a small bit of stem, leaves removed and placed in an empty vase. This completes the drying process and keeps also their shape.

Land Lotus Flower

Botanical Name : *Centella asiatica* (Linn.) *urban*

Family Name : Umbelliferae, Apiaceae

Common Name : Gotu kola, Kola, Indian Pennywort, Indian Ginseng, Bua Bok

Land lotus is a small annual, slender, creeping herb. It has long stalked green reniform leaves with rounded apices that have smooth texture with palmately netted veins. The flowers are pinkish to red in color enclosed in two green bracts. Fruits are small and flattened. It has been called the "miracle elixirs of life" because legend says that an ancient Chinese herbalist lived for more than 200 years as a result of using the herb.

Uses : Dried Land Lotus is versatile in its uses. The medicinal part is the dried aerial parts. The dried leaf powder is used as a medicine mixed in grape juice for controlling heart rate and blood pressure. It is also used for the treatment of leprosy. For this 1 gm of dried leaf powder is filled in capsules. It is also sold in dried form in the form of liquids. It is available in teas as dried herbs. People having insomnia are advised to take dried land lotus in a cup of water for about 4 to 6 weeks. An infusion of dried flower can also be used and it also serves as a tincture. It is also used in making fragrant potpourris.

Nigella Flowers

Botanical Name	:	*Nigella sativa*
Common Names	:	Black Seed, Black Cumin, Love-in-the-Mist
Cultivation	:	Full sun in average soil with excellent drainage, monthly fertilizer application is a must
Colors	:	Green seed pods with burgundy stripes
Regions	:	India, Egypt, Greece, Turkey

Drying Nigella

Nigella can be easily dried. Nigella is cut when pods reach maximum size and color. Nigella pods range from deep green to deep burgundy, with stripes in between. The first few pods are picked with as long a stem as possible, one at a time. The plants flower profusely. So when most of the flowers become seed pods, the entire plant is pulled and the stems cut. Then a rubber band is wrapped around a small bundle of stems and then hanged upside down for drying in a dark, airy place until dry.

Uses of Nigella

"Love in the mist" as nigella is commonly referred to, is used exclusively for dried arrangements. These flowers are the best to add texture to any dried flower arrangement. The delicate, purple striped pods are used in several arrangements for an airy effect. Dry roasted nigella seeds flavor curries, vegetables and pulses. They are also used to repel insects from clothes and are given to treat intestinal worms and nerve defects to reduce flatulence, and induce sweating. Dried pods are sniffed to restore a lost sense of smell.

Marigold Flowers

Botanical Name	:	*Tagetes patula, T. erecta*
Common Name	:	Aurora Yellow Fire
Origin	:	South Europe
Cultivation	:	Partial shade to full sun
Colors	:	Red, Orange, Bright Yellow
Bloom Time	:	June
Uses	:	As winter decoration and medicinal use
Regions	:	Desert areas of India, South America, Australia

Cultivation of Marigold

Marigolds are sown directly into a finely prepared seed bed. The crop requires supplemental irrigation. Both overhead and furrow irrigation should be available. The soil surface must be kept moist for uniform germination and emergence. Irrigation by sprinkler is advised to enhance seedling establishment and minimize soil crusting. Phosphorus is required to promote flowering. Nitrogen should be applied two or three times, during growing season.

Drying Marigold

The flowers are gathered only in fine weather, in the morning. Either whole flower heads or petals can be dried. To dry petals, the ray florets are used and need to be dried quickly in the shade where there is good movement of warm air. The florets are spread on sheets of paper, loosely without touching each other. If they are not loosely spread they will become discolored. To dry whole flowers any of the methods can be employed.

Uses of Marigold

Dried marigold is used for wonderful and exclusive dried arrangements. They are also used in broths, drinks and as a comforter of the heart. A plaster made with the dry flowers in powder applied to the breast strengthens and succors the heart in fevers.

Millet Flowers

Botanical Name : *Setaria macrochaeta*
Cultivation : Full sun in average soil.
Colors : Green
Uses : A nice dried flower texture, used as a dried flower accent
Regions : Dry regions of India, South Africa

The plant is an annual by habit with erect culms; 30-60 cm long. inflorescence comprising only a few terminal and axillary spikelets; subtended by a spatheole; enclosed. pedicels filiform; 0.2-0.4 mm long; glabrous.

Drying of Millet Flower

Millet is cut while they are still green. They are collected, tied together and hanged upside down in dark, dry, open, airy place. After they are completely dried they are collected together and stored in a place away from sunlight.

Uses of Millet Flower

They are used widely in dried flower arrangements. Mixing grains and grasses together form a stunning "textured" dried flower arrangement. They make wonderful arrangements in the form of wreaths, baskets, bouquets and as centerpiece to enhance the decor of any home. Flowering stems are dried and used for infusions and medicinal preparations.

Orchid Flowers

Botanical Name : *Cymbidium* sp.
Cultivation : Full sun with well-drained soil
Colors : All colors except blue
Uses : In dried arrangements and medicinal uses
Regions : Hilly regions of India, USA, Philippines

Cultivation of Orchid

Orchid Flowers have two patterns of growth *viz.*, monopodial and sympodial. In monopodials, the main stem grows indefinitely by one main bud in the center. The stem grows straight up and wide aerial roots generally grow from where stem and leaf meet at the base of the stem. They need lots of moisture and usually need 50 to 70% shade. They generally do well in very warm lightly shaded conditions. In sympodials, a main axis is present, but it grows mostly horizontally. They have several swollen shoots that are connected to each other via the main axis.

Drying of Orchid

Orchid Flowers are picked up when almost in full bloom. These flowers are dried by freeze drying. The flowers after cutting are hydrated and treated with a specific solution. They are then frozen for about 14 days in a freeze dry machine. The machine slowly extracts moisture from fresh flowers at extremely low temperatures. Each flower is then treated to protect them from fading and re-absorbing moisture.

Uses of Orchid Flowers

The beauty, perfume, and medicinal uses have been known since last few years. They are used in dried arrangements in wreaths, sachets and topiaries. The dried roots are used as antidepressants. They are used in cough syrups. Dried leaves have many applications in sweet or savory dishes and teas. Dried orchids are widely used in for medicinal and decorative purposes.

Oregano

Botanical Name	:	*Origanum vulgare*
Common Name	:	Wild Marjoram, Joy of the Mountain, Thyme
Origin	:	Native to Mediterranean Region
Cultivation	:	Full sun with well-drained soil
Colors	:	Pink

Uses	:	A beautiful dried flower
Regions	:	Warm regions of India, Europe, Greek.

Cultivation of Oregano

Oregano can be propagated by root divisions in spring or by seed. It requires full sun with well-drained soil to grow. Seeds germinate in about 4 days. Seeds should not be covered with soil. It is susceptible to root rot and fungal disease.

Drying of Oregano

Oregano can be collected when the plant is 6 to 8 inches high. Alternatively, the tops are snipped off. Stems are cut 4 to 5 inches from the ground. Flowers are harvested in summer or early autumn, just before full flowering. The washed branches are tied into small bundles and hanged upside down in a warm, dry, well-ventilated area out of direct sunlight. Good air circulation is a must so that the herbs do not get mouldy. They can be air dried outdoors, but better color and flavor retention usually results from drying indoors.

Uses of Oregano

Dried oregano is a beautiful dried flower with beautiful natural colors. Dried oregano is also sprinkled on top of a tomato and onion salad drizzled with some olive oil for a quick side dish. The dried herb contains several constituents, including volatile oil, such as carvacrol, thymol, and borneol. The thymol and carvacrol contents are responsible for its antimicrobial and antifungal effects. Dried leaf of oregano can be made into a tea by steeping 1 to 2 tablespoons in hot water for ten minutes.

Phalaris

Botanical Name	:	*Phalaris aquatica*
Common Names	:	Canary Grass, Plumosus
Colors	:	Light Green
Cultivation	:	Full sun in average soil
Bloom Time	:	June and July

Uses : A dried flower texture for any dried flower arrangement

Regions : Irrigated area of India, North America, Europe

Growing phalaris is very easy. It is a grass that can tolerate floods and extreme drought.

Drying of Phalaris

Phalaris is cut before maturity to maintain structure and prevent shattering. After cutting, they are tied in small bunches and then hanged upside down in dry, dark, open, airy place. Phalaris can also be dried in a microwave.

Uses of Phalaris

Phalaris is a nice dried flower texture for any dried flower arrangement. This crop takes up dyes easily. It is mixed with grains for a stunning textured dried flower arrangement. It is used to create accent in an arrangement. They have several medicinal uses. They are used for fighting pains and fatigue.

Purple Majesty

Botanical Name : *Pennisetum glaucum*

Common Name : Ornamental Millet

Cultivation : Full sun in average soil

Colors : Purple

Uses : A nice dried flower texture and offers a unique natural color

Regions : Karnataka, Warmer regions of India, Mexico, Australia

Drying Purple Majesty

Simple air-drying method is used for drying purple majesty. They are cut while flower plumes are still dense, collected and tied in small bunches. They are then hanged upside down in dry, dark, open, airy place.

Uses of Purple Majesty

Dried purple majesty is a stunning addition to any dried arrangement. They have many decorative uses indoor and outdoor. They are used in topiaries, swags, wreaths and baskets. These are air dried to produce tasty purple ruffles basil for use all winter. The dried roots are used in perfumery, sachet powders and to flavor dentifrices and toothpowders. The dried powder is said to act as a good snuff, useful to excite sneezing to relieve cases of congested headache.

Rudbeckia Flower

Botanical Name	:	*Rudbeckia hirta*
Common Names	:	Black-Eyed Susan, Orange Coneflower
Origin	:	USA
Cultivation	:	Partial shade to full sun
Colors	:	Bright Yellow, Orange, Red
Bloom Time	:	June to September
Uses	:	Ornamental uses
Regions	:	Northern regions of India, North America

Cultivation of Rudbeckia

Rudbeckia needs a partially shaded to full sunlight position for good flowering healthy plants. Plants which grow in areas of low light tends to be stunted and produce few flowers. Flowers are watered thoroughly and allowed to dry moderately between watering. Plants that are kept slightly moist seem to perform better. Established plants are quite drought tolerant. Media should have good drainage properties. Mixing compost into poor or impoverished soil improves their drainage.

Drying of Rudbeckia

Flowers are selected just before they are fully opened. The best method of drying is by embedding in silica gel either in open or in microwave. Flowers take about two and a half minute to dry.

Uses of Rudbeckia

Dried rudbeckia is widely used for ornamental and decorative purposes. Dried delights preserve these flowers for use in arrangements, wreaths and swags. Tea from dried leaves has been used as a kidney stimulant. Dried root tea is used to treat cold. They are also used to heal burns.

Sea Holly

Botanical Name	:	*Eryngium planum*
Meaning	:	Thistle
Cultivation	:	Full sun in well-drained soil
Colors	:	Lavender, Violet
Bloom Time	:	Mid Summer
Uses	:	An interesting natural blue accent
Regions	:	Mountainous regions of India, USA

Cultivation of Sea Holly

Sea holly is cultivated in a place with full sun having well-drained and preferably a gravel soil. The roots will not grow large or as fleshy if grown in a gravel soil. Plenty of sun is essential for all varieties. The best time to transplant the roots is in autumn, when the leaves decay. The young roots are much better to transplant. They will require no more culture than to be kept free from weeds.

Drying Sea Holly

Sea holly is picked up when the flowers are in full bloom. They are collected and bunches are made out of it. These bunches are tied together and hanged upside down in a dark and well-ventilated place. After drying, flowers are untied and carefully stored in a place away from direct sunlight.

Uses of Sea Holly

Dried sea holly is excellent flower with a natural blue accent. They are used as a decorative in winter bouquets. The dried root of the plant is powdered to aid tissue regeneration. Dried petals are added to healing mixtures and pot pourries.

Sweet Annie

Botanical Name	:	*Artemisia annua*
Common Names	:	Wormwood, Sweet Annie
Cultivation	:	Full sun in average soil
Colors	:	Green, Light Brown
Uses	:	Is an excellent dried flower filler
Regions	:	Coastal regions of India, Africa, South America

Sweet Annie can be easily grown from seed. Plants grow best in sunny location, but does not need any special soil. They have feathery green foliage. They hold their color very well when dried.

Drying Sweet Annie

Sweet Annie is dried for everlasting beauty. These flowers are cut before flowers open. After cutting them they are collected and made into small bunches. Later they are hanged upside down in dry, dark, open, airy place. After they are dried they are stored in a covered container that allows air circulation. They are made more durable by using a spray-on fixer which coat and strengthen brittle stems.

Uses of Sweet Annie

Flowers are dried for excellent decor. Sweet annie is a excellent dried flower filler. They are used extensively as dried flower wreath and swag base. Roots are fragrant and can be dried and used as potpourri. They are also used as insecticides.

Verbena Flower

Botanical Name	:	*Glandularia bipinnatifida*
Common Name	:	Mock vervains
Origin	:	South America
Cultivation	:	Full sun in well-drained soil

Colors	:	Pink, Rose, Purple, Red, Mauve
Bloom Time	:	Mid Summer
Uses	:	Decorative and medicinal uses
Regions	:	Assam, West Indian Islands, Kansas, Kentucky

Cultivation of Verbena

Verbena needs a sunny position, protected from strong winds and frost. Soil should be light, calciferous and well drained. If the soil is poor, the herb gets more hardy. If the hanging branches are cut away, it will be more fine and compact.

Drying Verbena

Verbena are cut when they are just about to mature. Flowers are dried by hanging. They can be hanged inside paper bags to keep dust off. There should be holes in the paper bags to let the air in and keep out moisture. They can also be dried by spreading them on trays. Microwave drying can also be used to dry verbena. These flowers are spread on a mesh rack and place in a slow oven set from 100 degrees to 125 degrees F. At times verbena are also press dried.

Uses of Verbena

Dried verbena is most commonly used in dried flower arrangements. It also has some medicinal uses. It is used as a tonic, astringent, diuretic, diaphoretic, galactagogue. It also makes a refreshing skin toner. It is an excellent addition to potpourri. A few leaves placed in a stationary box lend pleasant fragrance to the pot pourrii.

Zinnias

Botanical Name	:	*Zinnia elegans*
Cultivation	:	Full sun in well drained soil
Colors	:	Pink, Red, Orange, White, Yellow
Uses	:	As a dried flower filler in dried arrangements
Regions	:	The Himalayas, Arid regions of Rajasthan, North America and Rocky mountain

Cultivation of Zinnia

Seeds of zinnia are sown in a good potting mix and germinated at room temperature for 7 to 14 days. Zinnias are heat tolerant and make great cut flowers.

Drying Zinnia

Air drying or drying using a dessicant is best. The dessicant used is a mixture of 1 part borax and 1 part cornmeal. Cover the bottom of a box with a thin layer of the mixture and then place the flower face up. Sift on more of the mixture and place the box in a dark, dry place until blooms are completely dehydrated. Zinnias are also dried using sand.

Uses of Zinnia

Dried zinnia is widely used in dried arrangements. These flowers provide distinctive indoor decoration. Dried stems are used as filler in the bouquets. It is also used as dried ornamental and in potpourri. They are used in tomato dishes, pasta sauces, vegetables and soups. The aromatic intensity of the plant increases after drying.

These flowers are treasured by gardeners as heirloom plants and have proven themselves to be hardy enough to withstand weather and climate extremes. They are treasured in their dried form for beautiful and exotic dried flower arrangements.

Summary

Annuals are plants that complete their life cycle within the space of one year. Annual flowers are versatile, sturdy and come in different forms and colours. So, they are a favourite type of flower which is preserved. These flowers are dried and stored for use in several beautiful dry flower arrangements. Some annual flowers suitable for dry flower production are amaranthus, celosia, dahlia, gomphrena, helichrysum, hydrangea, marigold, orchids and chinese lanterns. The method of drying differs for each plant.

Question and Answers

1. What is an annual plant?
2. Name a few annuals suitable for drying and describe the most suitable method of drying them?

Chapter-15

Perennial Plants for Dry Flower Making

Perennial Flowers

Perennial flowers have a life span of more than one year and usually bloom during a specific time period every year during the growing season. These flowers are hardy and herbaceous. They need less maintenance, less water, and fewer pesticides than annuals. Perennials are easily used as ground covers, mixed with annuals, grown in containers and used as accents or specimen plants. Many perennials are short bloomers and are best mixed with others that bloom at different times or included with other landscape plants. These flowers are treasured by gardeners as heirloom plants and have proven themselves to be hardy enough to withstand weather and climate extremes. They are treasured in their dried form for beautiful and exotic dried flower arrangements.

Perennial Flowers

Baby's Breath
Bells of Ireland
Blue Globe Thistle
Carnation Flowers
Daisy Flowers
Larkspur
Lavender
Lily Flower
Monkshood
Pearly Everlasting

Flax
Hibiscus
Hollyhock
Iris
Lady's Mantle
Rose Flowers
Statice
Sunflower
Yarrow

Baby's Breath

Botanical Name	:	*Gypsophila paniculata*
Meaning	:	Pure Heart
Cultivation	:	Partial shade to full sun in rich, non-acidic soil
Colors	:	White, Purple
Uses	:	As the best dried flower filler
Regions	:	Coastal Regions of India (Western Ghats), Europe (Mountainous regions and farmlands)

It is a hardy annual able to withstand periods of short term drought. It bears white flowers forming dense delicate clusters. It prefers full sun in well drained soils. It cannot tolerate acid soils.

Drying Gypsophila

Gypsophila is a beautiful bushy plant with white clouds of flowers. These flowers are picked when about half of the florets are opened on the flowering stems. These are then collected and hanged upside down for drying in an open, airy place. These flowers may also be freeze dried. The flowers after cutting are hydrated and treated with a specific solution. They are then frozen for about 14 days in a freeze dry machine. The machine slowly extracts moisture from fresh flowers at extremely low temperatures. Moisture leaves the floral in shape and its size intact. Each flower is then sealed with a post treatment to protect them from fading and re-absorbing moisture.

Uses of Gypsophila

Dried gypsophila is used as the best dried flower filler. They serve as the most beautiful dried flowers in vibrant wreaths and swags.

Blue Globe Thistle

Botanical Name	:	*Echinops Ritro*
Common Name	:	Blue Glow, Vetches blue
Cultivation	:	Full sun in average soil
Colors	:	Purple Blue
Uses	:	Beautiful, striking, natural blue dried flower
Regions	:	Dry regions of India, Mexico

An easily grown hardy perennial for sun or partial shade with splendid, steel-blue, globular heads of flowers, excellent for flower arrangements. The metallic lustre of the flower heads keeps them decorative for a long time when dried. Globe Thistle has prickly foliage and produces blue, globe shaped flowers that have metallic appearance. They are drought tolerant and grow well in full sun with a light soil.

Drying of Blue Globe Thistle

The stems are cut when the flowers heads are just beginning to open, at the peak of blue. the prickly leaves are stripped off which allows for easier handling later. The stems are hanged upside down in dry, dark, open, airy place for drying.

Uses : Blue Globe Thistle forms a beautiful dried flower display with striking spherical shaped heads. They are commonly used for making exclusive wreaths, bouquets, swags and sachets.

Carnation Flowers

Botanical Name	:	*Dianthus caryophyllus*
Meaning	:	Divine Flower
Common Name	:	Purple Bouquet
Cultivation	:	Partial shade to full sun in well drained soil
Colors	:	Rose, Yellow, White, Red, Pink, Burgundy, Salmon
Uses	:	For ornamental and decorative purposes
Regions	:	Gujarat, South India, Coastal India, Washington, North Carolina

Cultivation of Carnation

Carnation grows well in a sunny position in most good soils but dislike acid soils. This is a plant that can tolerate salt laden air and strong winds of the coast.

Drying of Carnation

Carnation Flowers can be dried using air drying method but this method gives varied result. The best method to dry carnation is using silica gel. The use of silica gel retains more of natural color. The flowers are picked at its peak and embedded in the desiccants silica gel. For quick drying of carnations, they can also be dried by microwave drying method where they can be dried within 3 to 5 minutes.

Uses of Carnation

Dried carnations are widely used in dried arrangements. The flower heads are dried and used in potpourri, scented sachets and cosmetic products. Dried carnation flowers are also used for magical purposes and believed to add energy and power when used during a ritual and as incense.

Hollyhock Flowers

Botanical Name	:	*Althea rosea*
Common Names	:	Garden Hollyhock, Double Flowered
Cultivation	:	Full sun in fertile, well drained soil
Colors	:	Pink, Red, Yellow, White
Bloom Time	:	May and June
Uses	:	Decorative and medicinal uses
Regions	:	Indian Ocean Islands, North America, New Zealand

Cultivation of Hollyhock

The requisites for the production of a fine bloom of hollyhocks are a deep, rich, moist soil with perfect drainage. Good drainage is essential. They also require full exposure to air and light. They love sunshine and fresh air.

Drying of Hollyhocks

Hollyhocks are harvested when they are open and dried by using a desiccant. The flowers are placed in a cardboard box and the bottom is covered with desiccant. The flowers are placed on a layer of desiccant and the desiccant is carefully poured so that they cover the petals completely and only the stem shows. The lid is closed and the box placed in a warm, dry place. For quick drying a container should be used instead of a cardboard box. Microwave drying can also be done but the door should be kept open to check the flower color.

Uses of Hollyhocks

Dried hollyhocks are used in a plethora of dried arrangements and potpourri blend. They have a myriad of medicinal uses. They are used for several ailments.

Lily Flower

Botanical Name : *Lilium sp.*

Common Name : Massa

Cultivation : Sun to partial shade in well drained soil

Colors : Red, Orange

Uses : Decorative purposes

Regions : Temperate and Sub Tropical regions of Asia, China, North Africa

Cultivation of Lily

Lilies prefer well drained soil which is kept moderately moist, but avoid over watering. They need full sun but can tolerate partial shade. The unopened blossoms may fail to bloom if there is insufficient moisture in the air.

Drying of Lily

Lily can be picked in prime condition and air dried for best results. Lilies are also dried by freeze drying. The flowers after cutting are hydrated and treated with a specific solution. A protective coating is applied over the flowers and then they are frozen for about 10 to 14 days. The machine slowly extracts moisture from fresh flowers at extremely low temperatures. Moisture helps maintain the floral shape and size intact. Each flower is then sealed with a post treatment to protect them from fading and re-absorbing moisture. Lilies can also be dried by press drying.

Uses of Lily

Dried lily flowers are called golden needles. Dried lilies are used in exotic dried flower arrangement. They are wonderfully showy and fragrant. They have several other uses as well. They are used in potpourris for scent and color. It is also used in wine for flavor and color.

Monkshood

Botanical Name : *Aconitum napellus*

Common Name : Wolf bane

Cultivation : Best in semi-shaded location with rich soil

Colors : Blue-White, Cream, White, Deep Blue

Uses : A wonderful dried tall line flower

Regions : Great Plains of India, Middle Europe, Siberia

Drying of Monkshood

Monkshood are cut just before they start to bloom. Usually air drying gives good results.

Uses of Monkshood

Dried monkshood is a wonderful dried tall line flower and works well in vertical arrangements. Dried foliage is used in exotic herbal wreaths. Dried root is used as analgesic, anodyne, anti rheumatic, diaphoretic, diuretic and in homeopathy as a sedative. They are also used to make herbal tea. Dried flowers give color and flavor to beer, soups and stews.

Rose Flowers

Botanical Name : *Rosa sp.*

Cultivation : Full sun to partial shade in well drained soil.

Colors : Red, Orange, Pink, Yellow, White

Uses : Decorative, ornamental and medicinal uses

Regions : Widely distributed in a range of climatic and geographic conditions.

Cultivation of Rose Flowers

Rose flowers prefer full sun, but will be successful in partial shade, especially if the sun is during afternoon. The best soil should have good drainage, slightly acid and medium in texture (neither sandy nor clayey).

Drying of Rose Flowers

Rose Flowers can be Dried by Three Methods

Air-Drying : The roses should be allowed to open to more of a full bloom as opposed to drying them when they are still buds. The bottom parts of the stems are usually stripped off for easier handling. The stems should be staggered so that they are not clumped together. A light mist of spray on each rose is given and then hanged upside down in a warm and well ventilated place to dry.

Desiccant Drying : The roses are placed on a layer of silica gel such that the blooms face up. The blooms are then covered carefully with more of desiccant and the container sealed till drying is complete. Microwave Drying - Long stems are cut for this type of drying. The flowers are laid on moisture free paper and the oven adjusted to lowest setting. The paper is placed in the oven and checked periodically to prevent burning.

Uses of Rose Flowers

Rose Flowers are the most pre eminent and beautiful dried flower for adding joy, love and grace to flower arrangement. Dried flower wreaths are embellished with the dried roses. Dried rose petals are often more accepted by churches and reception halls. They can be used to enhance vases, baskets, plaques and shadow boxes. They also may be used as wall decorations or as decorations on gift boxes. They are used to make delightful potpourri. Red roses, when dried turn a rich burgundy color and look lovely in a ginger jar, candy dish, or antique canister. These potpourris can be added to the stuffing of pillows or are hung in the pomanders in a closet. Small sachets can be made out of it and can be given as gifts to friends and loved ones.

One can also make confetti cones and tea with dried petals. These petals have an essential oil in it which is healing and soothing. The dried hips of wild rose are especially high in vitamin C and have long been used to prevent scurvy. Infusion of dried rose petals are used for headaches and taken after meals to aid digestion. They make a good anti-inflammatory remedy for the eyes or any other inflamed area of the body.

Statice Flowers

Botanical Name : *Limonium latifolium*
Common Names : Sea Lavender
Cultivation : Full sun in well drained soil
Colors : Blue, Violet, Lavender
Bloom Time : July to September
Uses : A useful dried flower offering an array of bright natural colors
Regions : Southwest India, West Indies, Africa, South Florida, Tropical Regions

Cultivation of Statice

Statice grows well in full sun in well drained soil. They prefer sandy soils and are fairly drought tolerant.

Drying of Statice

Statice should be picked before it blooms fully. The florets continue to open as they dry. Hanging upside down in a dark room with good air circulation ensures drying in 7 to 10 days. If it dries too much resulting in brittleness, it can be misted with plain water to bring back its flexibility. The colors will last for years if the flowers are kept out of direct sunlight.

Uses of Statice

Statice is a useful dried flower offering an array of natural colors. They are preserved for use in dried arrangements, wreaths, swags, baskets and bouquets. They are used for making homemade potpourris. They have some medicinal uses also. The dried and powdered root is applied to old ulcers or piles.

Summary

Perennial flowers have a life span of more than one year and usually bloom during a specific time period every year during the growing season. These flowers are hardy and herbaceous. They need less maintenance, less water, and fewer pesticides than annuals. Perennials are easily used as ground covers, mixed with annuals, grown in containers and used as accents or specimen plants. Many perennials are short bloomers and are best mixed with others that bloom at different times or included with other landscape plants. These flowers are treasured by gardeners as heirloom plants and have proven themselves to be hardy enough to withstand weather and climate extremes. They are treasured in their dried form for beautiful and exotic dried flower arrangements. Some perennials suitable for dry flower production are baby's breath, carnation, daisy, larkspur, lavender, hibiscus, hollyhock, rose, statice and sunflower. The method of drying differs for each plant.

Question and Answers

1. What is a perennial plant?
2. Name a few perennials suitable for drying and describe the most suitable method of drying them.

Chapter-16

Dried Grasses

Dried Grasses

Grasses are easy to collect to dry and use. When the seed-heads have formed, the grass is cut off just above the soil using a sharp knife. They are allowed to lie in a thin layer in the sun for a day or two, turning them over a few times. They should not be allowed to get too damp. After they are dried they are put into a plethora of uses. Bunches of dried grasses can make a wonderful addition to any home or office.

Barley

Botanical Name : *Hordeum vulgare*

Cultivation : Full sun in average soil

Barley grass is one of the few green grasses. Astounding amounts of vitamins and minerals are found in green barley leaves. The leaves have an ability to absorb nutrients from the soil. It also contains a glucan, a fiber reported to reduce cholestrol level. The root contains the alkaloid hordenine, which stimulates peripheral blood circulation. Uses : Barley adds a nice texture to any dried flower arrangement.

Shama Grass

Botanical Name : *Echinochloa colonum*

Common Names : Billion Dollar Grass

Shama grass is a coarse annual grass cultivated in Japan and southeastern Asia for its edible seeds and for forage. Stem is erect spreading or lying horizontally on the ground and bending upwards but rooting from nodes in contact with the soil. These are used in their dried form to make a blend of potpourris and are used to flavor certain recipes. They can be used beautifully in dried arrangements.

Strip Grass

Strip grass, also called filter strip, filters runoff and removes contaminants before they can reach water bodies or water resources. They bear natural green flowers and can be dyed to a range of different colors. Dried strip grass is also used to weave into baskets.

Quaking Grass

Botanical Name : *Briza maxima*

Common Name : Marguerite Daisy

Cultivation : Full sun in very well drained soil

Uses : Is a nice dried flower texture

Regions : Northern regions of India, USA.

Drying of Quaking Grass

Quaking grass is harvested soon after heading, well before seed pods turn tan. If there is noticeable shattering of spikelets during handling the crop, this indicates that the plant was cut late. After cutting, the flowers have to be collected and hanged upside down in dry, dark, open, airy place.

Uses of Quaking Grass

Dried quaking grass is used as a nice dried flower texture. It is mixed with twigs and dried leaves to make wonderful arrangements. They are used as an accent in several dried arrangements in the form of bouquets and baskets. They enhance the decor and give a wonderful look to the interiors. The roots can be dried and ground into flour for making breads. They have many uses in traditional medicines, especially for kidney ailments.

Reed Spadix

Botanical Name	:	*Acorus calamus*
Common Names	:	Sweet Flag
Origin	:	South Eastern United States
Regions	:	Swamps, marshes, riverbanks
Height	:	1-4 feet
Flower size	:	Tiny flowers on a 2-4 inches spadix
Flower color	:	Yellow and green
Flowering time	:	May to August

Reed spadix is an herbaceous, perennial, aromatic, flag liek plant, flourishing in wet situations. The leaves are long, radical and sword like, erect, bright green but reddish at the base. The stalk is triangular, giving off from one side a spadix bearing thickly crowded perfect greenish flowers. It grows in muddy places, in swamps and meadows along streams and the borders of lakes, in nearly all part of India and Central Asia.

Uses

The dried reed spadix is a carminative, slightly tonic and forms a useful adjunct to other tonics and stimulants. They have lemony scent and are used to flavor candy. They can be used as substitue for vanilla pods. They can also be stored in dry foods to prevent infestation by weevils. They are a wonderful addition to potpourris.

Broom Grass

Botanical Name	:	*Thysanolaena maxima*
Common Name	:	Broom Corn, Broomsedge
Origin	:	North America
Regions	:	Bangladesh, North East India

Broom grass is a tall, reed like rhizomatous grass. Clumps are solid, smooth and rounded and as thick as a fore finger. Leaves are large and broad. Flowers are borne in a dense bushy, foxtail-like terminal inflorescence. It is an ornamental grass which is grown for its attractive foliage.

Dried broom grass creates attractive natural arrangements. It makes wonderful arrangements in the form of wreaths, baskets, bouquets, and as centerpiece to enhance the decor of any home. These are used as a local astringent and homeostatic.

Wheat Grass

Botanical Name : *Triticum aestivum*

Uses : Wheat grass is used as a dried filler to make beautiful dried flower arrangements. Along with its usage in the dried arrangement, it is also a potent component in the medicinal field. Dried wheat grass has a high protein content and is used for the treatment of protein deficiency and other related problems

Summary

Grasses are easy to collect to dry and use. When the seed-heads have formed, the grass is cut off just above the soil using a sharp knife. They are allowed to lie in a thin layer in the sun for a day or two, turning them over a few times. They should not be allowed to get too damp. After they are dried they are put into a plethora of uses. Bunches of dried grasses can make a wonderful addition to any home or office. Some important grass species used as dry flower are barley grass, shama grass, strip grass, quaking grass, reed spadix, dried broom grass and wheat grass.

Questions and Answers

1. What is the special feature of grasses as dried flower material?
2. Name a few species that can be dried successfully and explain their drying procedure.

Chapter-17

Drying Foliage

Dried Leaves

Leaves are dried or preserved to capture the beauty of spring and hold it for a long time. Dried leaves are beautiful when used in any decoration. There are several preserving techniques that help to enhance the prized collection of leaves well. Drying leaves is an age old tradition. Leaves were dried to preserve them. The leaves were picked for their medicinal qualities, fragrance, or to be used as spices and for their decorative value. Leaves can be preserved by various ways:

a) Placing leaf between two pieces of wax paper and then sealing the wax paper together by the use of iron.

b) In a solution of glycerin and wate,r the leaves are placed for almost 6 days to dry. The leaves will slowly absorb the solution and become soft and flexible for many years.

Drying in microwave oven by laying them between two paper towels.

Arrow Leaves

Botanical Name : *Arrow arum*

Common Name : Tuckahoe, Green Arrow Arum, Duck Weed

Bloom Time : May to July

Arrow leaves, named for their shape are narrow and leathery with unequal basal lobes. They are magenta to maroon beneath with small green spathes and are borne in a cluster. The leaves are fleshy and have veins. Dried arrow leaf is used in dried arrangements.

Cobra Leaves

Botanical Name : *Darlingtonia californica*

Regions : California, Oregon

The cobra plant is a herbaceous perennial consisting of a rhizome, with fibrous roots. Tubular pitcher leaves arise directly from the rhizome, forming a rosette. The hood of the pitcher is well developed to form a dome, with the pitcher opening facing downward. The pitcher lid is modified into a two-lobed, fishtail like appendage projecting downward from the front edge of the opening. The leaves twist as they grow. The dainty flowers have five petals, which hang from the base of the dangling flower. Cobra leaves are used in making beautiful decorations. They are used to make a attractive cover of any book and are widely used in craft works to make paper bags, show pieces, folders, etc. They can also be used as a bookmark. They are used for decorative as well as ornamental purposes.

Kanchan / Folding Leaves

Botanical Name: *Bauhinia varigata*

Kanchan leaves are also called folding leaves. These are the leaves that can be folded. In the dried form kanchan leaves are used as feed and manure. They are also used as flavoring agents and sweetners. They serve as a good antiseptic and disinfectant. They form an important ingredient of potpourris.

Palm Leaves

Palms are uniquely tropical trees. Their range is restricted mostly to the tropics. Palms are more closely related to grasses and orchids than they are to other kinds of trees. Palm tree leaves are used as thatch for roofs and some palm trees are grown as ornamentals. The most common used palm tree is coconut palm. The coconut palm, (*Cocos nucifera*), is edible in both fresh and dried form. Dried coconut leaves are used as a source of oil. When it is boiled in water, the oil is released and skimmed off for later use. Dried leaves are also used to make ropes. The wild date, palmyra, doum palms (*Hyphoene thebaica*), raffia palms (*Raphia sp.*) and nipa palm (*Nypa fruticans*) are most commonly used for flavoring wine. The simplest fuel usages of palms are exemplified by the burning of dry palm leaves. Dried leaves of nut palms is carved into beads and buttons. The palmyra palm has been used for making strings for musical instruments. Dried palm leaves are also used for various decorative purposes.

Spoon Leaves

Botanical Name : *Drosera intermedia*

Common Name : Sundew

Origin : East Africa

This carnivorous plant is found growing in muddy edges of ponds, bogs and rivers. It is a small herbaceous perennial, aquatic plant with short and slender fibrous root, from which grows the leaves. Their leaves are orbicular on long stalks, depressed, lying flat on ground and have on upper surface long red viscid hairs. These leaves possess the exclusive ability to dry naturally in insufficient sunlight. It can also be dried by air-drying and by placing the leaves in any desiccating material for few weeks.

Uses

Dried spoon leaves are used in dried flower arrangement. Dried spoon leaves are used by the best chefs to flavor foods before and after cooking. It is of greatest value in the treatment of spasmodic chest conditions such as whooping cough, bronchial and asthma. It is also prescribed for gastric problems and can be used as infusions.

Summary

Leaves are dried or preserved to capture the beauty of spring and hold it for a long time. Dried leaves are beautiful and essential for any decoration as a background. There are several preserving techniques that help to enhance the prized collection of leaves well. Leaves can be preserved by placing leaf between two pieces of wax paper and then sealing the wax paper together by the use of iron or by placing the leaves in a solution of glycerin and water or by drying in microwave oven by laying them between two paper towels. Some important leaves that preserve well are arrow leaves, cobra plant leaves, kanchan leaves and palm tree leaves.

Questions and Answers

1. What are the techniques used to preserve leaves?
2. Name a few palm leaves used for drying and mention the method of drying used.

Chapter-18

Dried Petals

Dried Petals

Many types of dried petals are used for making pot pourii eg., land lotus petals, rose petals, mahogany petals, lily petals and lotus petals. Flowers are dried for everlasting beauty. They add an elegant touch to the room and are dried for everlasting beauty. Preserving blooms is a wonderful way to preserve memories and enjoy long lasting splashes of color. Many flower parts are used in dried form. A vibrant and thrilling range of dried petals are used to embellish greeting cards, folders, books, and other personal belongings. They are also used as a potent component of cosmetics including perfumes. Using wide imaginations one can enhance the beauty of plain picture frames, mirrors and little boxes with dried petals. Each petal has its own beauty, which makes for gorgeous decorations.

Land Lotus Petals

Botanical Name : *Centella asiatica* (Linn.) Urban.

Common Name : Gotu kola, Kola, Indian pennywort, Indian ginseng, Horse-hoof grass, Pegaga, Mandookaparni, Tiger herbal, Spadeleaf, Tono

Land lotus is a small, annual, slender, creeping herb with long stalked green reniform leaves. They bear small pink or red flowers. Each flower is partly enclosed in 2 green bracts.

Dried land lotus petals are beautiful for decoration. They are placed in a flat container filled with water. The water in the container can be colored with dye or turmeric. In yellow shade the beautiful dried petal looks amazing. This can be used as a centerpiece and this centerpiece is enchanting and inviting. These dried petals can also be used for crafting projects. Extracts of the dried petals are added to cosmetic masks and creams to increase collagen and firm the skin.

Rose Petals

Botanical Name : *Rosa sp*

Colors : Red, Orange, Pink, Yellow, White

Regions : Indian Ocean Regions, South Africa, Germany, Widely distributed in a range of climatic and geographic conditions.

The rose with its fragrance and beauty is very often considered the queen of flowers. Have you ever wondered what gives the rose its delicious scent? Tiny perfume glands are present on the petals, which gives them a sweet scent. These tiny glands can be seen under the lens of a powerful microscope. Rose Flowers that have thick petals produce a stronger scent than those of thinner petals.

Uses of Rose Petals

Dried rose petals have been traditionally used in medicine, perfume, cosmetics, potpourri, and skin care. They are widely and beautifully used in cookery to make jam, syrup, beverages, deserts, etc. The syrup made of dried petals is a good source of vitamin C and is particularly good for children and small babies, as it is not too acidic for their immature stomachs.

Lotus Petals

Botanical Name : *Nelumbo nucifera*

Common Names : Nagau, Baino, Seroja, Bua-luang, Teratai, Kamal, Hasu, Lilian

Lotus is an aquatic perennial originally from Indonesia and India. This graceful flowering water plant, grown since ancient times, has a place of honour in the history of three civilizations: Egypt, China and India. It is considered to be a symbol of beauty, purity and perfection. On the culinary side, every part of the plant is used in cuisines. It is multi petalled pink flower. Dried lotus petals are ground and infused with water to make a fragrant tea. The petals steeped in wine or tea has a calming effect. The petals are placed in soups and are used as garnish. Petals are mashed into a paste and used as filling for cakes. The lotus petals are widely used in the preparation of pot pourii.

Lily Petals

Botanical Name : *Lilium*

Common Name : Massa

Cultivation : Sun to partial shade in well drained soil

Colors : Red-Orange

The petals of lily are called tiger lily petals. They are used in potpourri.

Angle Wing

Botanical Name : *Microcentrum retinerve*

Common Names : Lesser Angle-Wing Katydid, Angular-Winged Katydid

Angle-wing is a shrubby plant and needs good light and well drained soil. The leaves are dark green in color. Angle-wing is a wispy floral that look great in all kinds of projects and is an artful addition to any home. It is translucent to light forms.

Summary

Dried petals are used for making pot pourii eg., land lotus petals, rose petals, mahogany petals, lily petals and lotus petals. They are used to decorate greeting cards, folders, books, and other personal belongings. They are also used as a potent component of cosmetics including perfumes. They are also used to decorate picture frames, mirrors and gift boxes.

Questions and Answers

1. Name a few flower petals which can be dried and what products can be prepared from such dried petals?

Chapter-19

Dried Fruits and Pods

Dried Fruits and Pods

The enchanting beauty and usage of fresh pods or fruits need not end up with the season. They are dried and preserved for years. A colorful array of choice dried fruits are very popular in dried arrangements. There are numerous dried fruits and pods used in potpourri and dried flower arrangements. There is a wide variety of fruits that are successfully used in their dried form and are dried by different techniques depending on their types and requirements. Well-dried pods should be packed in clean, dry gunny bags and stored ensuring protection from dampness after removing the extraneous matters.

Arjun Pods

Arjun pods are small tiny pods that can be used as beautiful raw material for decoration. They are colored with a splash of colors and can be used for crafting show pieces. They are popularly used for making potpourris. They make fragrant blend of potpourris and spread a sweet aroma wherever placed.

Badam Nut

Botanical Name : *Prunus amygdalus*

Common Names : Almond, Mandel, Sweet Almond

Origin : Native to South West Asia and North Africa

Drying

Almonds are harvested when the hulls have cracked open. Traditionally the hulls are removed and the nuts dried in sun until the kernels rattle in the shells. These days it is also dried by placing it in a desiccating material for few days.

Uses

Sweet almond oil is obtained from the dried kernels of almond. The oil has no scent and is a great emollient, softening and conditioning skin and also helps skin to balance its loss and absorption of moisture. The oil is well known to relieve itchy, dry and inflamed skin. Dried badam nut are imported from India. These are used in decorations. They are placed in their natural color with approximately 19" long on stems in a vase. They can also be made more beautiful by painting them gold or silver.

Belani Lacquered

Common Name : Bael Nuts

Botanical name : *Aegle marmelos*

The bael tree has its origin from Thailand and India. The tree is deciduous with trifoliate aromatic leaves and grows to a height of 6 to 8m. The size of the mature fruit varies from 550 g to more than 3 kg. It has natural splotchy skin. It is large, hard, round ball about 2 to 3 inches in diameter. It may or may not have a stem. It is one of the oldest used ingredients for potpourri. The bael nuts can be varnished and used in dry flower decorations.

Bell Cup

Bell cups are large, hard and 2 to 3 inches in diameter. The most fascinating thing about bell cups is their shape of a cup. They look beautiful as decoratives. They can be mounted on a cane stem. They can be easily dyed using varied colors, specially golden, saffrron, and yellow. They are available as natural, stonewashed or copper glitter. Bell cups with fragrant seasonal greens and carnations provide an impressive dried arrangement that will enhance any christmas table.

Blue Pine

Botanical Name : *Pinus wallichiana*

Common Names : Himalayan Pine, Butan Pine

Growth Rate : Moderate to slow

Site Requirements : Sun to partial shade, well drained soil, good wind tolerance, tolerates seasonal wet soil and drought

Texture : Medium

Form : Graceful, upright, loosely and broadly pyramidal when young, often retains lower branches, wide spreading with pendulous branches. Blue pine is an evergreen tree growing to 25m by 10m at a fast rate. It is in leaf all year, in flower from May to June, and the seeds ripen September to October. The flowers are monoecious and are pollinated by wind. Flowers are cylindrical cones, pendulous and turn brown with age. It comprises of a five-needle pine with a bluish hue on its foliage. Branches are whorled and leaves arranged in spirals. The plant prefers light, medium and heavy soils, requires well-drained soil and can grow heavy clay and nutritionally poor soils. It cannot grow in shade. Dried blue pine forms beautiful dried flower arrangements.

Casuarina Pods

Botanical Name : *Casuarina equisetifolia*

Common Name : She Oak, Jungli saru

Color : Red-Brown

Casuarinas can grow as small shrub to tall trees. The leaf is long, thin, pine like needle shaped. The male flowers are usually the tiny flowers that grow in to long spikes at the end of the stems. The female flowers grow into a nut like fruit. Casuarinas are easily raised from the seed and will grow under a wide range of conditions. Dried casurina pods are widely used for victorian centerpiece and exclusive swags to add charm and beauty to any place. They are one of the most commonly used decoratives. They are one of the premium scented botanicals used for potpourris.

Cones

Cones is a general term used for the fruit of a pine tree. A pine cone is shaped like an egg or a cone. It has scales on them. Pine cones usually have spiny points at the tip of each scale. There are a number of species falling under the pine category. All of them are used in some or the other way. Dried Pine Cones can be used to prepare a variety of decorative items like wreath, bird feeders, scented pine cones, pine cone kindler, pine cone cushion, pine cone easter decorations, thanks giving turkey, christmas trees, angel ornaments, mini bird's nest, corn cob magnet, pine cone pony, pine cone snow people and pine cone skier craft.

Some Important Species of Pine

Lodgepole Pine (*Pinus contorta*): This also known as shore pine. It is a common tree in western North America. Like all pines, it is evergreen. The fruit of this pine tree is small, egg shaped with a prickle at the end of each scale. These fruits may remain close on the tree for years. There are three sub species of *Pinus contorta.*

- *Pinus contorta* subsp. *contorta* var.*contorta* (Shore pine) and var.*bolanderi* (Mendocino shore palm)
- *Pinus contorta* subsp. *murrayana* (Tamarack Pine or Sierra Lodgepole Pine)
- *Pinus contorta* subsp. *latifolia* (Lodgepole Pine)

The species name *contorta* arises from the twisted, bent pines found in the coastal area. *Pinus contorta* is occasionally known under several English names: Black Pine, Scrub Pine, and Coast Pine. *P. contorta* subsp. *latifolia* will hybridize with the closely related Jack Pine - *Pinus banksiana*.

Ponderosa Pine (*Pinus ponderosa*): The fruit of this tree is a egg shaped cone. Each scale has a straight, stiff prickle that sticks out. *Pinus ponderosa*, commonly known as the Ponderosa Pine, Bull Pine, Blackjack Pine, or Western Yellow Pine, is a widespread and variable pine native to western North America. Like most western pines, the ponderosa is associated with mountainous topography. The Ponderosa Pine has a very distinct bark. Unlike most conifers, it has an orange bark, with black lining the crevasses, where the bark "splits". There are four different taxa of Ponderosa Pine, with differing botanical characters and adapted to different climatic conditions. These are:

- *Pinus ponderosa* sub sp. *ponderosa*
- *Pinus ponderosa* sub sp. *scopulorum*
- *Pinus brachyptera*
- *Pinus benthamiana*

Jeffrey Pine (*Pinus jeffreyi*): The fruit is a large, woody cone with scales. Each scale having a curved prickle that curves inward. It is a high altitude species and is tolerant of serpentine soils. The Jeffrey

Pine is a large tree, reaching 25 to 40 m (82 to 130 ft) tall, rarely up to 53 m (174 ft) tall, though smaller when growing at or near tree line.

Knob cone Pine (*Pinus attenuata*): The fruit is a woody cone with swollen, knob like bumps on one side. Cones may remain closed on the tree for many years. The tree grows in mild climates on poor soils. The Knobcone Pine crown is usually conical with a straight trunk. It reaches heights of 8–24 meters, however; it can be a shrub on especially poor sites.

Western White Pine (*Pinus monticola*): The fruit is a woody cone that is slender and curved. Cone scales are thin and often curve up on the end. Fruit is a huge, woody cone and cone scales are thick and straight. It is the state tree of Idaho, and is sometimes known as the Idaho Pine.

Limber Pine (*Pinus flexilis*): Fruit is a woody cone with thick cone scales having no prickles. *Pinus flexilis* is typically a high-elevation pine. In Limber Pine, the cones are 6-centimetre (2.4 inches)-12-centimetre (4.7 inches) long where the species overlap, green when immature, and open to release the seeds; the scales are not fragile. Limber Pines usually have intact old cones lying under them.

White Bark Pine (*Pinus albicaulis*): Fruit is a small, woody cone that is nearly round. They have thick scales with no prickles. They remain on the tree even when mature. Seeds do not have wings. They are a rich source of food for animals. The cones are 4-centimeter (1.6 in)-7-centimeter (2.8 in) long, dark purple when immature, and do not open on drying, but are fragile and are pulled apart by birds to release the seeds. White bark Pines almost never have intact old cones lying under them.

Lotus Pods

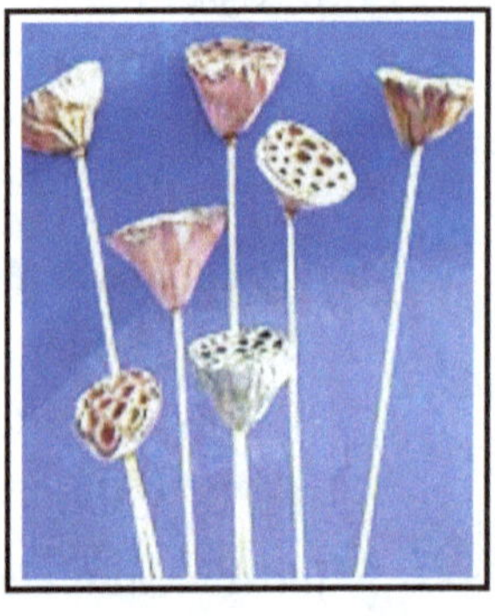

Botanical Name : *Nelumbinis nuciferae*

Lotus is an aquatic perennial flower. It has large flowers and leaves that sometimes grow up above the water. The plant's thorny stalk discourages fish from nibbling on it. The upper capsule or fleshy capsule of the lotus dries out at maturity and separates from the plant. Floating about, it scatters seed from the many perforated holes in its surface. Lotus pods are the fruits of the lotus flower. They are grey, bell shaped pods with flat tops, having numerous holes containing the pod's seeds. The average size of these pods runs between 1 to 2 inches in length.

Uses of Dried Lotus Pods

Dried lotus pods offer a great visual effect in any kind of potpourri. They can be placed in between dried arrangements to make wreaths, swags, and baskets. They make classic centerpiece by mixing them with dried materials. These pods can be colored with a range of colors to get more effective appearance and grace.

Pear Pods

Botanical Name	:	*Pyrus communis*
Common Names	:	Akras, Onkne, Apios

Pear pods have a woody outer skin that are about 2 ½ inches long. They are round in shape getting narrow at one end of the fruit. They are popularly used in making delightful and aromatic blend of potpourris.

Poppy Pods

Botanical Name	:	*Papaver sp.*
Cultivation	:	Full sun and sandy loam soil is best
Harvest	:	Cut poppy pods when the seed pod is fully mature and still slightly green
Drying	:	Hang upside down in dry, dark, open, airy place
Uses	:	They have a very good dried flower accent.
Regions	:	Mountaineous terrain with warm temperatures, high altitudes and dry soil

Poppy pods are the seed pods of the poppy plant. The poppy is a simple flower with a powerful drug as a product. It is a lovely annual flower with vividly bright leaves. The blooms come in vivid shades of red, pink, white and purple and each plant has one bloom. Flowers have large petals. A large seed appears after the poppy blooms. This seed pod is egg shaped and can be used in dried flower decorations.

Uses

Dried poppy pods are available in a range of different sizes, shapes and colors and are striking in any project. They are popular due to their aesthetic value and medicinal properties. They are beautifully used as dried flower accent. When arranging dried poppy pods, it is important to use strong base foam which will hold the dried pods upright and is much sturdier. Fragrant potpourris are made using dried pods.

Mango Fruit Pods

Botanical Name : *Mangifera indica*

Common Names : Mango

Regions : North East India, South and South East Asia, Burma

The mango tree is an evergreen tropical tree with a dense overhanging canopy. The fruits are sweet with thickish green to orange flesh around a large flat stone.

Pod Description

The mango pods are also known as Easter Nuts. These veined pods are an average of 1½ inches long. They are light brown in color with a rough surface. They are used in potpourri.

Bakuli Pods

Bakuli pods are tulip shaped wooden pods. They make a whistling sound in the wind. The dried pods are popular for pain relief, healing skin, nose and burns. It imparts physical energy and is also used in sachets and potpourris. It is of great medicinal value.

Bora Flower

Bora flowers are beautiful flowers with long pointed leaves and petals with a range of colors. Most commonly found colors of bora flowers are red and yellow. Alternatively, orange and deep yellow colors are also seen.

Uses

Bora flowers when dried, serve as a beautiful accessory to embellish every corner of your house. They are potent components of dried floral arrangements. They are perfect for use in potpourris and floral crafting. The dried petals of the flowers are used to decorate photo frames, book covers, scrapbooks, and endless number of creative gifts. Dried bora flowers when glued evenly spaced around a frame looks like a button from a clown suit. This makes a perfect primary colored frame for a child's room. They are also used to make scented gift boxes, candles scent and soaps.

Summary

A colorful array of choice dried fruits are very popular in dried arrangements. There are numerous dried fruits and pods used in potpourri and dried flower arrangements. There is a wide variety of fruits that are successfully used in their dried form and are dried by different techniques depending on their types and requirements. Well-dried pods should be packed in clean, dry gunny bags and stored ensuring protection from dampness after removing the extraneous matters. Arjun pods, almonds, bael, bell cups, pine cones, casuarinas, lotus and cones are commonly used dried material.

Questions and Answers

1. Name a few dried fruit pods used in dry flower arrangements.

Chapter-20

Artificial Flowers

Artificial flowers are a great source of handmade decorations. Beautiful handmade artificial flowers are suitable for banquets, weddings or to decorate any room in the home. They are generally made of paper or cloth and in a variety of colors. They come in either a bouquet form or can be ordered as individual stems. There are different kinds of artificial flowers including silk flower, latex flower, flowers made from natural leaves and embroidery flower.

20.1 Types of Artificial Flowers

Latex Flower : Latex Flowers can easily be shaped into almost any design as required because the stems are wired and hand wrapped.

They have a latex coating over natural parchment and so realistic in appearance. They can have a multitude of color variations.

Silk Flowers : Silk flowers are stunningly realistic looking flowers that are made from silk. They make an excellent, economic and aesthetic alternative to the real flowers.

Embroidery Flowers : These are wonderful flowers that are embroidered or made with beads. They utilize wires and embroidery floss. They are light weight and give the look of satin.

20.2 Uses of Artificial Flowers

Artificial flowers make a great gift, brightening up anyone's home, yard even hospital room. Artificial plants and artificial floral arrangements can give your room an elegant finishing touch. They are used in various styles with a complete line of colors to make a thrilling decorative piece. These flowers are outstandingly elegant and are infinitely suited to apparel and jewellery applications.

Decorative Shapes of Artifical Flowers:

Palm	Canes	Sola	Others
Palm Cap	Cane Coils	Sola	Coco Six Finger
Palm Ring	Cane Curls	Flowers	Coco Spiral
Palm Spear	Cane Spring	Sola Balls	Kamboi Spring
Palm Flowers	Cane Cone	Sola Rose	Roman Candle
Palm Cord Ball	Cane	Sola	Twig Cone
Palm Cones	Peacock	Sunflower	Twig Spring
		Sola Fans	Vine Candle/ Lantern
		Sola	Grass Ball
		Baskets	Coco Cap/Flower
		Sola Chips	Coco Chip/Petal

Palm Caps

Palm caps are open pods of the palm trees. The seeds are released from the pods and the pods form a great decorative piece used in several dried arrangements. They are colored with a splash of colors and make dramatic arrangements. They can be used as an accent or can even be placed in a glass vase with some colorful pebbles in it.

Their stately beauty enhances the most formal and elegant setting. They can be whimsical when placed on christmas tree full of lights.

Coco Cap / Flowers

Coco flowers are remarkably life like artificial flowers and are an elegant addition to any decor. They look as natural in the living room as they do outdoors. They add beauty and elegance to weddings, banquets and other affairs. They can be customized like live flower arrangements to fit any themes.

Sola Flowers

Sola flowers are handmade flowers from tapioca wood. It is a soft wood which can be easily strung together with a needle to make natural garlands and rose balls for decoration. Sola wood offers an exquisite range of floral treat to enhance the decor. These flowers are

Sola balls
(Tapioca soft wood peels covered over a plastic ball)

Natural Sola Tips
(1.5" to 2.5")

intricately designed and impart elegance par excellence. These flowers come in a black base that adds to their beauty. They are ideal as gifts and interior decoration accessory. They add a delicate touch to dried arrangements. They form great masterpiece and arranging these flowers can be a great art.

Cane Coils

Cane coils are long canes with a coiled structure at one end. They look beautiful in their original form and can be used for decoration in rooms and offices. They can be used as an accent in any floral arrangement. They can be painted to match the interior. They can be placed in deep vases or containers. The long stem can be cut to place them in a small decorative vase.

Sola Rose

Rose is itself a beautiful flower that adds joy and love to beauty. Sola rose is a beautiful flower to be gifted or use as an embellishment. To make sola rose cut the sola wood in the shape of the petals. Cut the petals in different sizes to show bloom. Take a small petal, roll it and tie it at the bottom with thread. Arrange few petals around it to complete a rose and tie all the petals together at the bottom. Sola wood is fragile so the petals should be rolled and tied carefully.

Sola Sunflower

Sola sunflower is the most vibrant types of sola decorative to enhance the decor of any interiors. To make sola sunflower sola wood is cut into the shape of the sunflower petals. A round piece from the sola wood is cut for the center of the flower.

One by one the petals are tied at the base of the round sola wood with a thread and arranged around the sola wood to complete a circle. Sola wood is fragile, so the petals should be tied carefully. The petals are colored yellow and the centre black for a natural appearance.

Summary

Artificial flowers are a great source of handmade decorations. Beautiful handmade artificial flowers are suitable for banquets, weddings or to decorate any room in the home. They are generally made of paper or cloth and in a variety of colors. They come in either a bouquet form or can be ordered as individual stems. There are different kinds of artificial flowers including silk flower, latex flower, natural leaves flower, natural look flower and embroidery flower. Some of the standard artificial flowers are available as Palm Caps, Rings, Spears, Flowers and Cones; Cane Coils, Curls and cones and materials made from sola wood.

Questions and Answers

1. What do you mean by sola flowers?
2. Name the shape of some artificial flowers.

Chapter-21

Principles of Floral Arrangement

Introduction

Floral arrangement is the art of selecting and organizing flowers and foliage according to the elements and principles of design in order to attain a pleasing and interesting design. Arranging flowers is an artistic activity and a way to express oneself creatively.

21.1 Flowers and Foliage

Successful floral arrangement begins with plant materials that have been properly handled and prepared. Beauty and good composition of an arrangement is not determined by the cost or rarity of plant materials used, but by the way they are selected, cared for and arranged. The plant material has to be handled carefully, handling only the stems. Properly dried, bleached and dyed flowers, foliage and accessories can be arranged following the principles for dry flower arrangement.

21.2 Containers

The size of the container should be in scale with the table or location where it will be used. The shape of the container should suit the arrangement to be made. Simple shapes are easiest to work with.

Ornate containers should be avoided as they may detract the viewer from the floral arrangement. Neutral colors like tan, brown, gray or greens are easy colors to harmonize with plant materials and with most backgrounds. Colored containers are available, but should be used with care so as not to detract from the arrangement. Containers come in many materials, such as glass, pottery clay, porcelain and metal. Containers can be thin and dainty, or thick and sturdy. For dry flower arrangements, a base is absolutely essential. The base can be a driftwood piece or any such natural or natural looking piece.

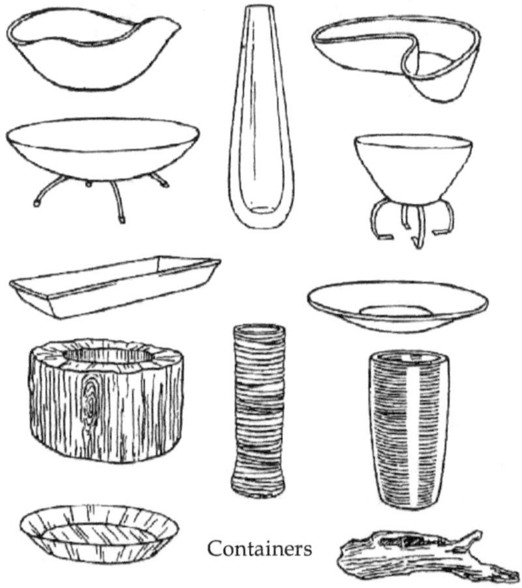

Containers

21.3 Holders

A good holder should give the freedom to position stems where they are wanted and hold them in place securely. A holder chosen depends on the style of arrangement planned, the plant materials to be used and the container to be used.

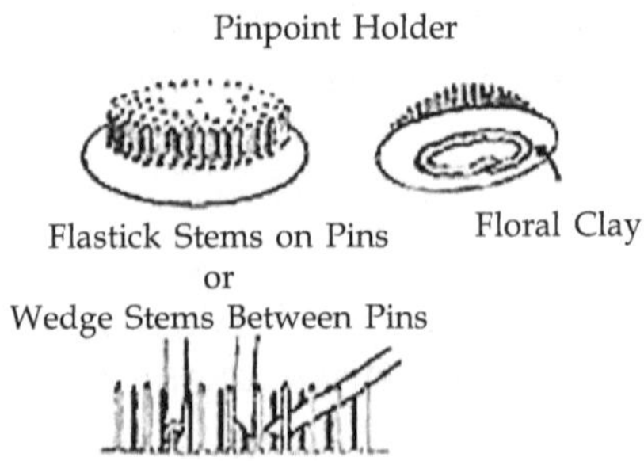

21.3.1 Pinpoint Holders

Pinpoint holders are most often used for line and line-mass arrangements in low bowls or shallow containers. Floral clay is used to fasten the holder to the base of the container. A heavy metal pinpoint holder is easiest to keep in place in the container. Stems are either stuck directly onto the pins or are wedged between the pins. Thin stems may be placed on a pinpoint holder by binding several stems together with a rubber band, string or floral tape.

21.3.2 Floral Foams

Floral foams are available in wet or dry forms; wet foams are used for arranging fresh plant materials, while dry floral foam is used for dried or silk floral arrangements. Wet and dry types of floral foams are not interchangeable, their intended use is specific. Floral foam is inexpensive and may be purchased from floral supply stores, arts & crafts stores or florists. Floral foam is best used for line-mass or mass arrangements, but may be used in some line arrangements also.

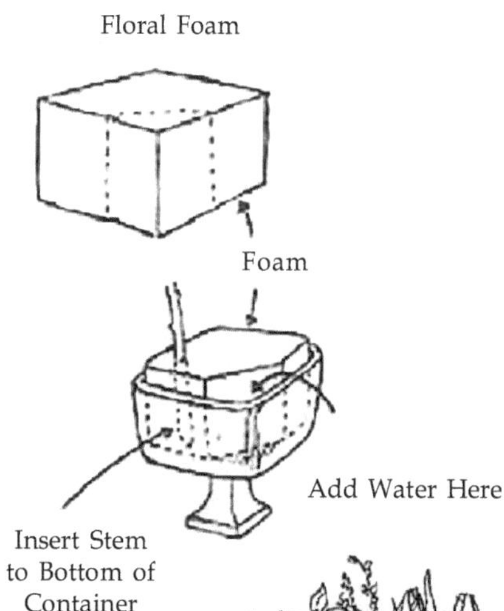

Floral tape may be used to secure the foam if needed.

21.4 Floral Arrangement Styles

Three geometric designs *viz.*, Line, Line-mass, and Mass, form the foundation of all floral designs.

21.4.1 Line Arrangements

Line arrangements are adaptations of Japanese styles. Linear pattern is dominant. Line arrangements are characterized by restraint in the quantity of plant materials used, with an emphasis on the beauty of individual blooms or foliage. Line arrangements are usually bold and dramatic, with importance placed on the contrast of form and texture, with an open silhouette. Often line arrangements will have three lines or placements. Most linear arrangements have asymmetrical balance. Typically, they are viewed from the front only. Allowing some material to extend toward the front or back of the arrangement develops depth.

24.4.2 Line-Mass Arrangements

Line-mass arrangements combine the strong line of Japanese styles with the massed effect of European designs. Line-mass arrangements have a clean, uncluttered look, with definite line, a well-defined mass, and plenty of open spaces. The dominant line is combined with a mass of plant material at the focal area, additional material is used to enhance and develop the linear shapes. This is a design combining unusual components and using more plant material than used in a Line arrangement. Contrast of texture, color and line are important features.

21.4.3 Mass Arrangements

Mass arrangements are adapted from European designs. They have a thick, full look, with a closed silhouette. Mass design uses more plant material than Line or Line-Mass designs; usually a large amount of plant material is used. The emphasis is on the whole colorful mass of flowers and foliage, rather than on individual components. Color is important in mass arrangements. Round or mass shapes usually dominate, but spiky or linear forms are good for triangular arrangements.

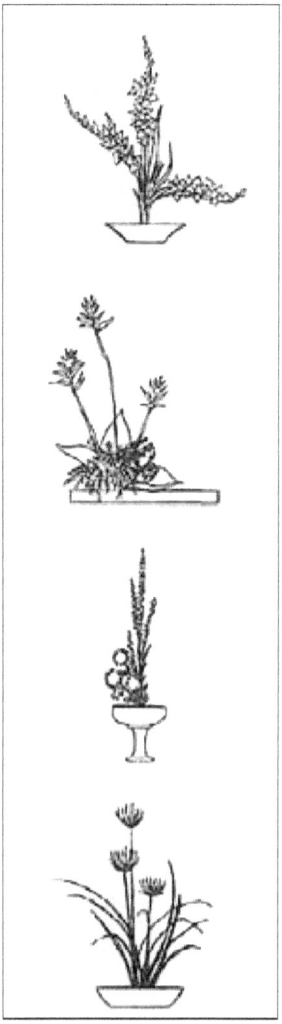

Line Arrangements

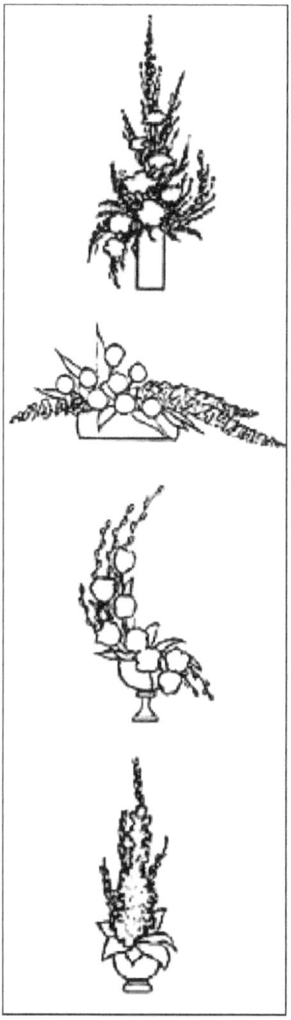

Line-Mass Arrangements

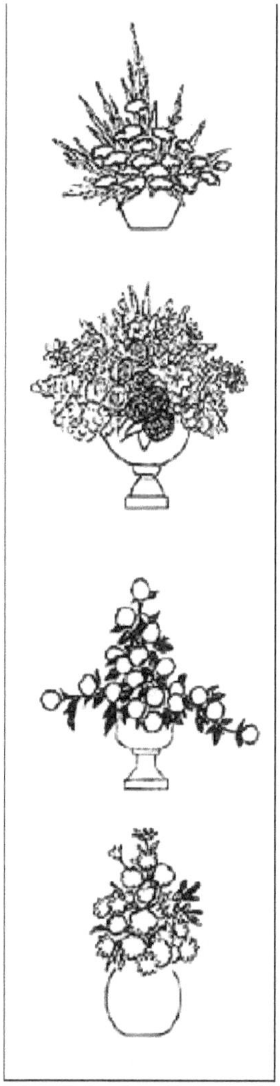

Mass Arrangements

Design

A floral arrangement is made whenever selected flowers and foliages are placed into a container according to a plan. Flowers casually placed in a container are attractive because of their beautiful color and shape, but the same flowers are even more appealing when arranged in a stylish way. Design is selecting and arranging of objects and deals with the form of their composition. Good floral design is the result of a well thought-out plan, with two aims in mind; order and beauty. For a good design, there must be a relationship between all of

Floral Arrangement Shapes

Crescent

Circle

Triangle

Oval

Mass Arrangement

Oval

Triangle

Triangle

Triangle

Diagonal Line

Simple Vertical

Spiral

its component parts; a feeling of security, naturalness, balance and simplicity. The floral arranger must have a basic idea, a mental picture of what they wish to create, where the arrangement will be placed, and the function it will serve. Most floral arrangements are made for a particular purpose or place. The arrangement should be suitable for its intended use. A good floral arrangement should be expressive of a theme or idea and of one's own personality. In order to arrange flowers in such a manner it is necessary to become acquainted with all the elements of good design. The elements of design are color, light, space, line, form, pattern, texture, and size. The principles of design are balance, dominance, contrast, rhythm, proportion, and scale. These same elements and principles of design apply to every art form. Some attributes of design are beauty, harmony and expression.

21.5 Elements of Design

21.5.1 Color

Colors have different effects on the viewer. Though use of colour in dried flower arrangements are highly restricted, the basic principles have to be applied especially when using dyed material. Effective use of color is important in floral arrangement design. Reds, oranges and yellows are generally bright and stimulating, and are considered warm colors. Warm colors tend to be advancing colors (to the eye), and have more visual weight than cool colors. Blues, greens, and violets are usually tranquil, peaceful and restful and are considered cool colors. Cool colors tend to be receding and have less visual weight than warm colors.

21.5.2 Light

Light affects design in many ways. Light may change apparent colors of materials, enhance form, texture, or depth. While light is a design element to be considered, in the fair setting, lighting is not under the control of the designer. Typically, lighting will be limited to artificial lighting in the form of overhead fluorescent lights. If the display area is in open-air, indirect natural light may be present during the day.

21.5.3 Space

Space describes the open area in around the design. Total space may be limited by restrictions imposed by the class schedule, the class description, fair rules, etc. Space also refers to the spaces within individual plant materials selected for use in the design. The designer also establishes spaces within the design through the placement of materials.

21.5.4 Line

The primary foundation of design is line, which creates a visual path for the eye to follow through a design. Lines also function to establish the structural framework or skeleton of a design. Line characteristics may be long or short, straight or curved, weak or strong, etc.

21.5.5 Form

Form describes the three-dimensional aspect of a design. A form may be closed - compact, massed, with few open spaces which is typical of a mass design. It may also be open, with spreading parts which produce spaces between the parts which are typical of a line or line-mass design. A closed form appears heavier than an open form of the same size. Form also applies to the shape of individual components of a design. Forms or shapes are generally put into three main groups according to their shape:

Spiky/Linear Shapes : Useful for line and line-mass arrangements. Often used to form the skeleton of an arrangement. Examples: gladiolus, iris, ornamental grasses, twigs and branches.

Mass/Rounded Shapes : Useful for mass or round arrangements. May be used to develop the focal point in line-mass arrangements or make up almost all of a mass arrangement. Examples: chrysanthemum, marigold, rose, zinnia, hosta leaf.

Spray or Filler Shapes : Useful for mass and line-mass arrangements. Used as a background fillers and as space fillers in mass arrangements. Examples: baby's breath, ageratum, ferns, asparagus leaves, parsley.

21.5.6 Pattern

Pattern is the design formed by solids and spaces. Individual components in a design have a pattern of their own; individual components are combined into an overall pattern.

21.5.7 Texture

Plant texture refers to the surface quality of plant material. Textural contrast and variety can add interest to a design. A plant's texture may be fuzzy, glossy, smooth, rough, etc. Texture also applies to the overall effect of the arrangement of the petals or florets; for example, a spray of baby's breath has an airy texture, a rose has a velvety texture, a zinnia has a dense texture.

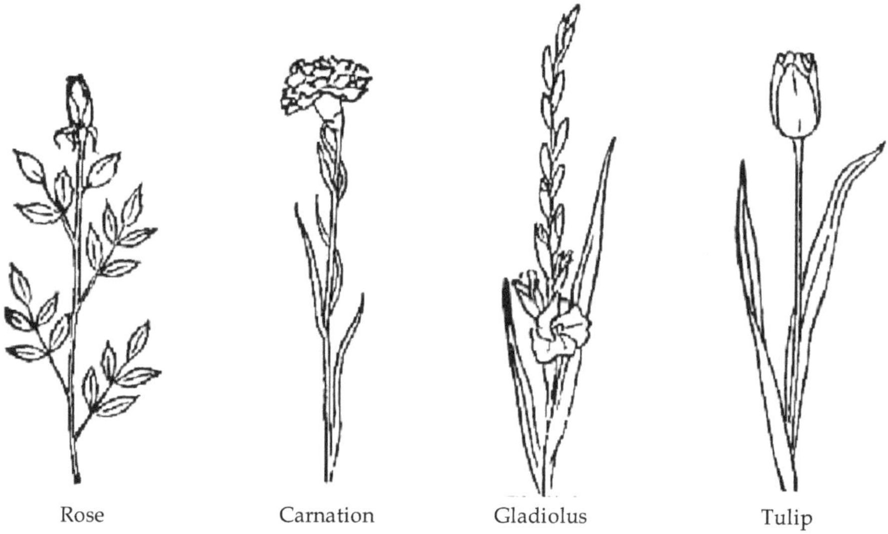

Spiky or Linear Forms

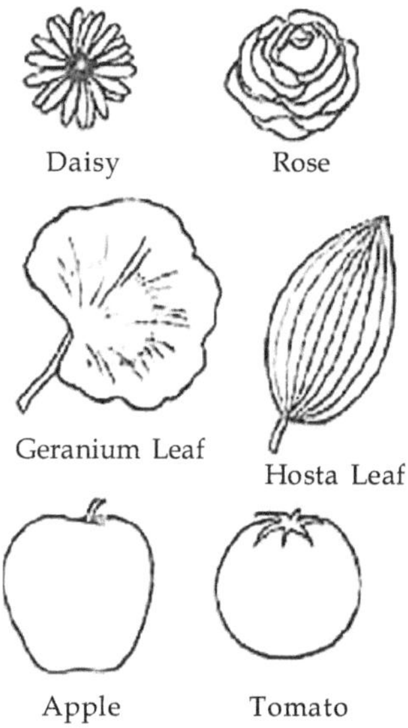

Mass or Rounded Shapes

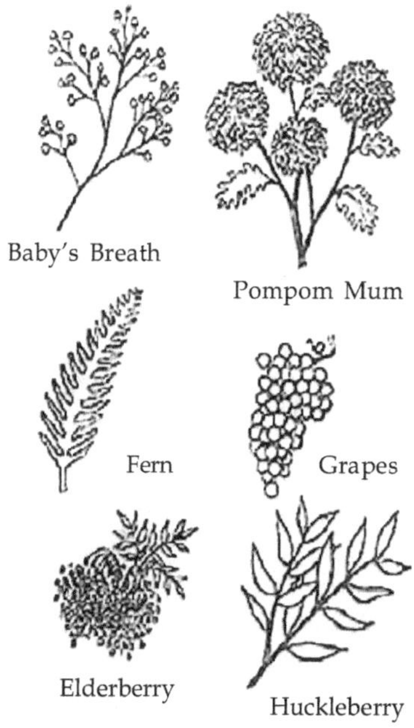

Spray or Filler Shapes

21.5.8 Size

Flowers and leaves of different size are selected. Usually buds and small flowers are used at the top and edges of an arrangement, while large leaves or fully opened flowers are placed low in the arrangement. A large element is often used as a focal point. In floral design, size also applies to apparent or visual size. A flower's size may be expressed as delicate, heavy or bold.

21.6 Attributes of Design

21.6.1 Expression

Expression is an attribute of design. Through the artistic and creative selection of components, a mood, feeling or idea may be expressed, and communicated to the viewer. For example daffodils and pussy willows suggest spring; cattails make one think of a swamp; sunflowers are casual and suggest summer; white roses are formal and elegant.

21.6.2 Harmony

Harmony is an attribute of design, and is the principle that produces a feeling of unity throughout the entire composition. To obtain harmony, all components must have something in common either in size, shape, texture, idea, time of flowering or color.

Harmony is not well expressed in this arrangement because such heavy, coarse flowers as zinnias have little in common with a spring flowering shrub.

Flowers and plant materials should be selected that have something in common such as flowering habit and blooming date. The narcissus harmonizes nicely with a spring flowering shrub.

21.6.3 Unity

Unity is an attribute of design, and is that quality that expresses congeniality, cooperation, and a spirit of working together as a unit. Unity is the fitting or bringing together of all of those principles and elements which go into making up the design or composition.

This ornate container in figure (1) detracts from the lilies because of its texture as well as unity.

Fig. 1 :

Fig. 2 :

The container used in figure (2) emphasizes the beauty of the lilies rather than detracting from them.

21.7 Principles of Design

21.7.1 Balance

Balance means that the finished arrangement does not fall over or look as if it will fall over. Consider the balance from top to bottom, side to side, and front to back. Balance is a feeling of rest, security or stability. There are two types of balance - symmetrical (or formal) balance and asymmetrical (or informal) balance.

Symmetrical balance (formal balance) means that one side of the arrangement is nearly a mirror image of the other. Symmetrical balance requires a repetition of flowers, buds, and foliage on both sides of the central axis with the two sides as nearly alike as possible. Asymmetrical balance (informal balance) means that the plant material and placement are different on each side of the central axis. Although the sides of the arrangement differ, they should have equal visual weight to maintain a sense of balance. Asymmetrical arrangements balance different flowers, buds, and foliage. The axis in informal balance does not need to be through the middle of the container.

In floral arrangement, formal balance is the most easy to execute, is dignified and impressive, but appears very mechanical. While formal balance is inspiring, its effectiveness depends greatly on one's ability to recognize values and arrange them in a restful manner.

In this arrangement, formal or symmetrical balance is illustrated. Both sides of the arrangement are composed of materials that are the same kind, form, size and color in equal amounts

In informal, or asymmetrical balance, the two sides may be distinctively different, but they have an equal weight such that the sides balance one another.

Examples of poor and good balance are shown with iris in these two arrangements. The same flowers have been used in both arrangements, but to better advantage in the one below.

21.7.2 Visual Balance

Visual balance depends upon placement of flowers and foliage. A form (flowers and foliage) with the same size and color used higher or further away from the axis will appear heavier than the same form (flower or foliage) used lower in the design and closer to the axis. Using larger or darker plant material just above the container creates visual stability. However, too much weight placed low in the design can make the design too bottom heavy.

21.7.3 Focal Point

That area or portion of a design which first attracts attention is the focal point of an arrangement. The focal point is usually placed low in the arrangement near the center, just above the container in such a way that it breaks the horizontal line of the container. Normally the focal point is composed of some massive material such as a large flower, bright color or something that will immediately attract attention, have weight and give stability to the arrangement.

This arrangement has a weak focal point because the darker flowers have not been grouped to form as strong a focal point.

This arrangement appears to be more stable and better balanced when the large, darker flowers are placed near the focal point.

21.7.4 Dominance, Accent or Emphasis

Dominance in floral arrangement is the greater force of a design element, such as more round forms, or more curved lines. Dominance may be achieved through the use of a dominant color, size, shape or texture, or by using larger forms or stronger colors. Accent may also accomplished by introducing sharp contrast in form, size or color. Arrangements should have a focal point to attract immediate attention or interest. Accent or emphasis is used to first attract attention to the most important feature in an arrangement (focal point) and from that point to every detail in order of its importance.

The focal point is definitely lacking here. There is a fixed line of demarcation between the container and flowering materials used.

By introducing the two leaves of sharp contrast in form and size, accent or emphasis has been placed where it is most needed.

21.7.5 Rhythm

Rhythm refers to movement. The swing or motion in an arrangement should create a dominant visual path, so that the eye is first attracted to the focal point and then carried throughout the entire arrangement. Rhythm in floral arrangements may be obtained through repetition of shapes, hues, line direction, etc. Rhythm may also be achieved through a gradual change or progression (gradation) in size, weight, texture, or color. A continuous line movement of flowers, stems, twigs, or color may also create rhythm.

In every arrangement there should be a feeling of swing or motion. Motion is lacking in the arrangement on the left. By rearranging a few materials, motion is quickly recaptured in the arrangement on the right.

21.7.6 Contrast

Contrast, or difference, is achieved by placing opposite or unlike elements together so as to emphasize or accentuate their differences. Contrast stimulates interest. Too much contrast divides attention.

21.7.7 Proportion

Components in the arrangement should be in good proportion to one another. Proportion is important in the use of color, texture or form. Consider the amount of plant material in relation to the container. Also consider the height of the arrangement in relation to its width.

Proportion

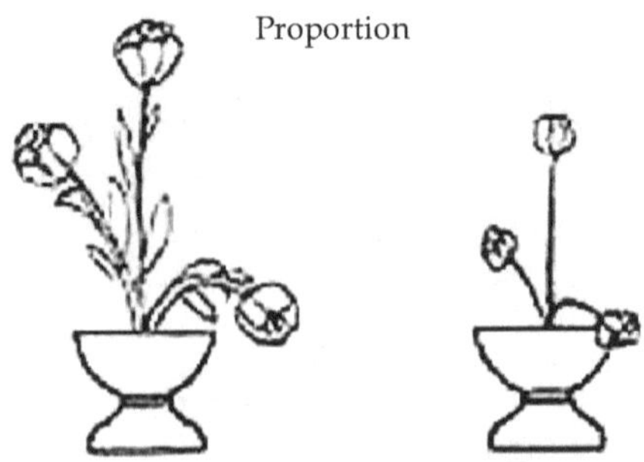

With regard to line, proportion is relative length. In regard to form or space, proportion deals with the relative area or volume. In regard to color, proportion is relative magnitude with the combined effect of color, hue value intensity and texture. Proportion in regard to shape would describe the number of round forms in relation to spike forms. Proportion of size would dictate the number of small flowers in relation to large flowers.

21.7.8 Scale

Scale is size relationship between flower and flower or between flower and container. When variation of size is too great or too small, components are out of scale. Small flowers will appear smaller when scattered among larger flowers. Better scale balance may be obtained if the smaller flowers are grouped, thus giving them more individuality. Placing small and large elements together accentuates the contrast between the elements. Scale requires special attention in miniature arrangements. Rhythm, contrast, proportion and scale.

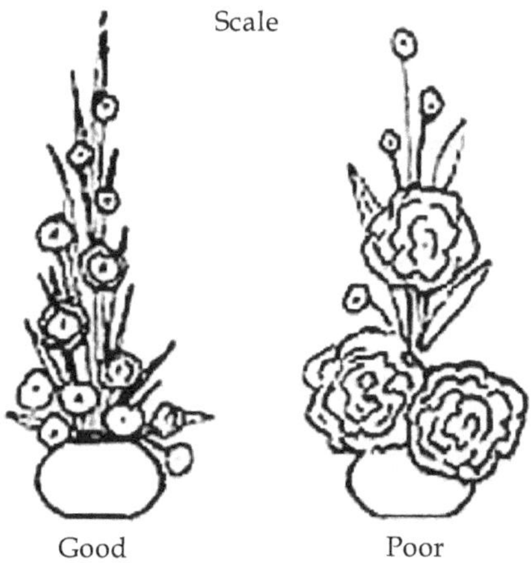

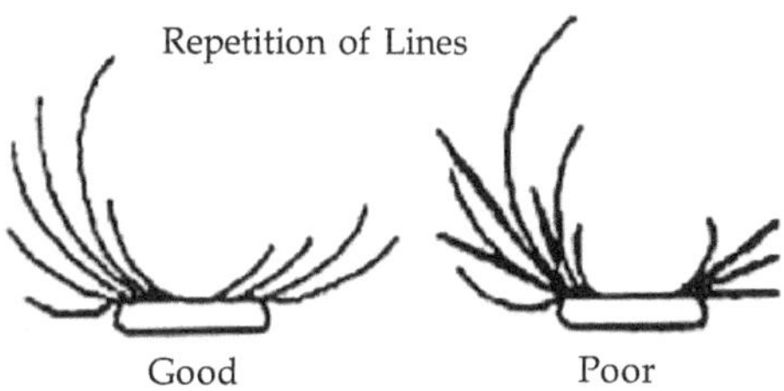

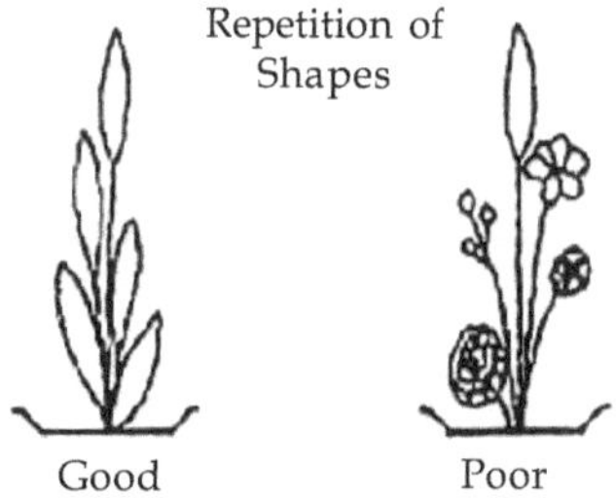

Summary

Floral arrangement is the art of selecting and organizing flowers and foliage according to the elements and principles of design in order to attain a pleasing and interesting design. Arranging flowers is an artistic activity and a way to express oneself creatively. Properly dried, bleached and dyed flowers, foliage and accessories should be arranged following the principles for dry flower arrangement. Any sober looking container can be used but for dry flower arrangements, a base is absolutely essential. The base can be a driftwood piece or any such natural or natural looking piece. Dry floral foam is used for dried or silk floral arrangements Three geometric designs *viz.*, Line, Line-mass, and Mass, form the foundation of all floral designs. Good floral design is the result of a well thought-out plan, with two aims in mind - order and beauty. For a good design, there must be a relationship between all of its component parts - a feeling of security, naturalness, balance and simplicity. The elements of design namely color, light, space, line, form, pattern, texture and size have to be considered for any arrangement. The various attributes of design like expression, harmony and unity and the principles of design like balance, visual, balance, focal point, rhythm, contrast, proportion and scale have to taken into consideration.

Questions and Answers

1. What are the elements of design?
2. What are the main pre requisites for dry flower arrangements? Describe.

Chapter-22

Value Added Dry Flower Products

Dry Flower Products

Dry flowers are aesthetically beautiful and they enhance the décor and liveliness of any room. They bring the fiery warmth and bounty of the season into the home with ease every fall. Much of fall's beauty lies in the glorious changing colors of autumn leaves. These flowers are dried to use them in a wide range of products. Dried floral products are widely used for decoration and for adding zest to a room. Apart from that they are also used as an essential component of many products. They are widely used to make handmade paper, lampshades, candle holders, jute bags, photo frames, boxes, books, wall quilts, topiary, cards and several gifts. The use of dry flowers in the making of these products enhances the appearance and beauty of these products. They look sophisticated when gifted to someone or used as a décor of your own home.

Dry Flower Products

1. Handmade Paper
2. Lamp shades
3. Boxes
4. Photo Frames
5. Candle Holders
6. Carry Bags

7. Wall Quilt
8. Books
9. Topiary
10. Jute Bags
11. Cards
12. Decoration

22.1 Potpourri

Potpourri is a mixture of dried, sweet-scented plant parts including flowers, leaves, seeds, stems and roots. The basis of the potpourri is the aromatic oils found within the plant. These oils are not confined to the flowers but they are at their peak at flowering time. So they should be harvested just as the plant begins to flower. A properly blended potpourri will last for months. Its uses are unlimited. Pot pourri are used to perfume the air with their continual fragrances, scent drawers and closets, make sweet smelling stationery, add a final touch to room decorations with attractive baskets or create personal gifts to share with friends. Apart from being widely used for fragrance and beauty, potpourris are also used to repel moths and protect woollen garments in storage. Beautiful scented potpourri is hard to beat as an attractive room accent. Use an assortment of dried buds and petals (roses, carnations, and lavender are always popular), and boost the aroma by adding cinnamon sticks, orange peel, and essential oils. Potpourri is a mixture of dried flower petals, leaves, roots and spices used to scent the air. A wealth of dried flowers and herbs are arranged to form potpourri to suit every occasion.

Crystal potpourri is a unique gel like crystal that can be simmered or left out to fragrance a room. Different combinations of herbs, seeds, flowers and spices create amazing aromas and textures. The materials used for this purpose should be resistant to mould, non toxic and able to withstand mechanical damage. The basis of pot pourri is the aromatic oils found within the plant (Geetha *et al.*, 2004). Boonsong, (2008) conducted studies on the improvement of aroma in Thai Pot pourri and dry flowers using starch film as an aroma fixative. The results of the study indicated that the starch mixed with essential oils of cloves constituted the best aroma fixative for potpourri and dry flower mixture.

22.1.1 List of Plants used for Potpourri Making

Amra Pods	Angle Wing	Bakuli Pods	Pine Pods
Buddha Nut	Cardamom Husk	Corn Grass	Cotton Flower
Eucalyptus Pods	Globe Amaranth	Hibiscus Pods	Luffa
Mahogany Pod	Maize	Nigella	Peepal Pod
Pine Needle	Screw Pod	Sponge Mushroom	Star Flower

22.1.2 Method of Making Mixed Potpourri

- Dried leaves and flower petals are placed into a large bowl.
- The bowl is gently stirred.
- Spices and seeds are mixed in small amounts to attain fragrance.
- A fixative such as orris root or gum benzoin is added.
- The ingredients of the bowl are tossed lightly and allowed to stand at room temperature for 5 to 6 weeks.
- The contents are shaken periodically.
- A piece of netting is cut to a desired length and glue applied to the inside bottom of a container. The netting is pushed into the container.
- The container is filled with potpourri and netting secured with a ribbon bow.
- Flowers glued at the bow add to the beauty.

22.2 Dry Flower Arrangements

Dried flower arrangements have been around for centuries. Dried floral arrangements could be seen in 19th century homes on pianos, sideboards and windowsills. These dried arrangements were found mainly during the winter months when garden flowers were absent. Traditional Victorian dry flower arrangements are opulent, colorful pieces. This style of dry flower arrangement can be modernized by varying styles and forms. A dynamic splash of vibrant and breathtaking dried flowers can be composed beautifully in several forms or arrangements. A selection of interior design baskets and other whimsical containers are used to enhance any interior design scheme. These arrangements form the best décor accents and gifts. They add a touch of elegance, enhanced beauty and cheerfulness to any office or home. They form a unique gift that adds an undeniable charm and brings a smile to just about any occasion. Moss, twigs, wheat and grasses also enhance the dried flower arrangement. A few silk flowers, ribbons or bows added to the arrangement ensure a more delicate and sophisticated look. There are a number of different dried floral arrangements including dry flower bouquet, mixed potpourri, bleach bouquet, basket bouquet, swag, garland, dry flower sachet, dry flower wreath, and box bouquet.

22.3 Dry Flower Sachet

A sachet is a small packet of perfumed powder used to scent clothes, as in trunks and closets. By using dried flowers a memorable gift can be made in the form of dry flower sachet. They can be a great gift for someone you love, spreading aroma of love everywhere. These are perfect for making drawers, closets, and other small areas smell sweet. Sachet can be of different shapes like heart shape, round or any other specific shape. The size depends on how many flowers are used. The materials needed to make a dry flower sachet are flowers, string, felt, glue, lace material, piece of lace, and perfume.

22.3.1 Method of Making Dry Flower Sachet

- The felt and lace material is cut into desired shape.
- Glue is put around the edge of the cut felt leaving a small section without glue.
- The lace is set on the glue and it is set aside to dry.
- After the glue gets dry, the petals of dried flowers are placed into the opening left in the felt.
- After stuffing the dried flowers, the opening is glued shut and laced appropriately.
- Once the glue is dry, dried flower sachet is ready.
- A small amount of perfume on the back of the felt ensures a sweet smelling sachet.

22.4 Dry Flower Wreath

A wreath is a ring or circlet of flowers, boughs, or leaves worn on the head, placed on a memorial, or hung as a decoration. By using dried flowers a colorful wreath can be created to hang on the front door or over the mantel. It adds elegance and liveliness to the beauty of your home.

22.4.1 Method of Making Dry Flower Wreath

- The stems from the dried flowers are cut. Flowers like lavender, hydrangea, sea foam, statice, and strawflowers are preferred.
- A floral wire is attached anywhere along the wire wreath frame.
- Several stems of foliage are selected and placed together in a bunch with the cut ends facing the same direction.
- A few dried flowers are laid on top of the bundle of foliage.
- The bundle is placed on the top of the frame where floral wire is attached.
- This bundle is held in place and the floral wire wrapped around the bundle and frame.
- Another bundle of foliage and flowers is gathered and laid so that the leaves overlap with the first bunch and cover the cut ends. The stems on both bunches should be facing the same direction.
- In the same way several bunch are overlapped and wired to the frame until a circle is completed.
- The last bundle is tucked under the first bundle attached to the wreath frame.
- Once completed, the wire is cut and the dried flower wreath is ready.

22.4.2 Hydrangea Wreath

Hydrangeas are big, glorious flowers produced in abundance on large garden shrubs. One of the wonderful things about them is that the flowers can be allowed to remain on the plant throughout the blooming season and then harvested in late summer. Grapevine twig wreaths and florist's wire can be used, circled around the top of the wreath a couple of times with a loop fashioned at the top. Strong garden twine can also be used. Flowers of the same height are used. Any discolored, badly faded or damaged flower petals are removed. When different colored hydrangeas are being used the overall color pattern should be decided before proceeding. A dry run is tried on a table, setting the florets down in a circle the same size as the wreath base. The florets are glued in place using a low-temperature hot glue gun. The flower stems are poked into the foam and then secured with a small amount of glue. Excess glue strings are usually easy to remove once they are dry. The same procedure is continued until the wreath is completely hidden by hydrangeas. At this point, sprigs of baby's breath can be added for a contrast in color and texture, but hydrangeas are classically elegant all on their own.

22.4.3 Grapevine or Twig Wreath

Grapevine or twig wreaths may use less hydrangea blooms, while also offering more individual style choice, because not the entire wreath needs to be covered. The most suited or preferred design is chosen by setting hydrangea florets in a circle of the same size as the wreath base, until the flower colors and contrasting spaces look balanced and right. The additional design elements like, a raffia or silk bow are glued on, before gluing on flowers on the wreath. Flowers are attached with a low-temperature glue gun. the Hydrangeas are interspersed with baby's breath, ivy, preserved leaves or other dried flowers such as roses or peonies or the hydrangeas alone look good enough.

22.5 Dry Flower Basket

This arrangement consists of a large array of dry flowers from sunflower, cockscomb, and poppy pods to artichokes, money plants, and gumballs, enhancing the décor of any home or office. Dried flower baskets are used for home decorating, heart throbbing gifts and house warming ideas. They customize any décor, brighten and personalize wall spaces and help create a warm, welcoming feeling in your home. These baskets can be in several forms including wall baskets, wedding flowers, preserved rose bridal baskets. These baskets are designed to accent seasonal change, such as autumn or spring; events like weddings, special messages like thanks giving, friendship, promotion, etc.

22.5.1 Method of Making Dry Flower Basket

- A large array of dried flowers are gathered and arranged in a basket of any shape. Hand woven Baskets are preferred.
- Dried flowers, foliage and preserved leaves are arranged to create a full, lovely display.
- The basket is decorated with ribbons and pinecones.

- One can also spray any good perfume to the arrangement.
- A beautiful basket of dried flowers is ready.

22.6 Wedding Decorations Flowers

In a natural arrangement of flowers; harvested grasses, dried flowers and lavender are tied up together simply with a ribbon to give the beauty of natural products. This natural dry floral arrangement adds a touch of simplicity with beauty to the décor of the house. They have a flat back and can be easily placed on any surface. It is a self standing arrangement composed of fragrant flowers.

22.6.1 Method of Dry Flower Arrangements

- Exotic dried flowers with long stems are selected.
- Any vase like floor standing or tall crystal vase can work for this arrangement.
- The flowers should be allowed to fall out naturally.
- This makes an elegant natural dried flower arrangement.

22.7 Box Bouquet

Flowers are well represented in the complete spectrum of fine, durable boxes. A large array of flowers can be arranged in a box giving a natural look to the flowers. This enhances the beauty and elegance of a home or office. This is a very charming arrangement, which is most commonly used as memorable gifts on weddings, anniversary, and birthdays.

22.7.1 Method of Making Box Bouquet

- The box is lined with a clean paper.
- A block of floral foam is cut to fit the box and the foam soaked in water for a few minutes.
- The dried flowers are collected and an asymmetrical arrangement created on the foam.

- Dried grasses and lavender is added to the arrangement.
- The foam is placed in the box and the box set on a decorative base.
- The flowers are sprayed and the arrangement is ready.

22.8 Basket Bouquet

A basket bouquet is a beautiful one-sided arrangement of dry flowers that add aesthetic beauty and charm to the décor of a house or an office. It is a lasting keepsake and is great for an intimate centerpiece. These baskets are sturdy and make a great gift for birthdays, get well, thanks, congratulations, or any occasion.

22.8.1 Method of Making Basket Bouquet

- A small brown paper bag or glass bottle is taken to hold the dry flowers.
- Lines are drawn on a bag to make upside down " T " with a thick horizontal bar.
- A cut is along the lines with decorative edge scissors.
- The bag is opened.
- A bouquet of dry flowers is arranged and the bouquet set inside the basket.
- The two halves of the handle are joined with a double-sided tape. A wire basket or wooden basket can also be used instead of paper basket.
- Foam is placed in the wire basket.
- The bouquet prepared out of dry flowers is set firmly on the foam, inside the wire or wooden basket.
- The basket is covered it with cellophane and decorated with matching ribbons.

22.9 Swags

Swags are green or dried material secured to some sort of flat frame. Dried floral swags can enhance a doorway or banister for holidays and events. They are made from glycerinized salal, white hydrangea, yarrow, globe thistle and globe amaranth and several other types attached to a common base. They form a wonderful gift to your near and dear ones.

22.9.1 Method

- Dried flowers are collected and the stems cut to a small length.
- A long string or twine is laid on a large flat surface.
- A loop is tied in one end of the string.
- A floral wire is attached to the loop end of the twine.
- Several foliage pieces are selected and placed together in a bunch, with the stems at one end.
- The bundles are placed at the loop end of the twine.
- The floral wire is wrapped around the stems and twined to secure them in place.
- The floral wire is wrapped around the bundle twice.
- Several dried flowers are laid on top of the bundle of foliage.
- Another bundle is prepared and placed such that the stems overlap with the first bundle.
- The wire is twisted tightly around the last bundle, wire knotted and strung together.
- The portions of wire not required are cut off.

22.10 Garland

Garland is a wreath or rope of flowers and leaves. A garland of flower buds and similar ingredients closely strung together and placed on the hair or on the upper forehead along the hairline. Such a garland should have flowers of two colors: red in the middle of the garland and the other colors flowers on the two sides.

Symbolizing romance, the elegant, versatile garland has become readily available since the advent of dry flowers. Dry flower garlands add beauty to any wedding ambience. They add interest and elegance to any doorway or entrance. They are widely used for ceremonies and receptions. They can be made quickly, stored easily and arranged in a variety of places.

22.10.1 Method to make a Garland

- A number of mini bouquets are made.
- The bouquet end is wired to the end of a long piece of vine, ivy, honeysuckle or rope.
- One bunch is placed halfway over the last bunch and wired with a gauge paddle wire.
- The wire is wrapped a couple of times and pulled tight.
- When the garland is sufficiently long, the first bunch is tied with the last bunch.
- The garland is placed on the mantel, or on any other place such as along a counter in the kitchen, on top of a mirror as part of the interior decoration.

22.11 Handmade Paper

Dried flowers make a wide display of colours. In making handmade paper, no chemicals and harmful dyes are used. It is made by using old cotton rags, dried flowers and other materials. Floral handmade paper are eco-friendly but expensive. They help in preserving nature. There is a plethora of colors and textures available for handmade paper products. Skilled craftsmen make these papers reflecting a blend of tradition and beauty.

Handmade Paper Products : From the handmade papers made of dried flowers, a number of other products are made including cards, gift boxes, notebooks, stationery, art portfolios, gift bags, etc.

22.11.1 Lamp Shades

Dried flowers are used to decorate lamp shades. These flowers look beautiful when light shining through them. An old lamp shade can be made exclusively beautiful with pressed leaves and flowers.

22.11.1.1 Making Dry Flower Lamp shades

The fabric on any old lamp shade is replaced with a plain new fabric of the desired color. The pressed leaves and flowers can be made into several designs, logos and pictures on the lamp shades. This pressed flower lamp shade rivals the beauty of floral arrangements on the product.

22.11.2 Boxes

Boxes are made from a wide range of materials with innovative, attractive designs, patterns made from dried flowers and leaves. Boxes make wonderful storage containers. They can serve as aesthetically beautiful gifts on weddings, birthdays and anniversaries. They add a personal touch to the gift.

22.11.2.1 Making a Dry Flower Box

Any wooden or paper box is painted with desirable colors and then dried flowers and leaves placed on them. The flowers are placed forming an exclusive design or pattern or some symbol. Writing names or messages with dried flowers and leaves can also customize these boxes. One can also blend dried flowers by applying glue to the box and spraying the powdered dried flowers unevenly on the box. A beautiful ribbon can be added to complete the arrangement.

22.11.3 Books

Books are a reservoir of information and wisdom and the only means to treasure your words, your knowledge. Books can be made beautiful decorated with dried flowers and leaves. Flowers are arranged on the wax paper into any design required. A small dab of glue onto the back of the flowers hold them in place. The flowers and wax paper is then, covered with a piece of clear contact paper. Gently press to make sure the contact paper adheres to all the flowers. This wax paper can be used to cover a book and make it beautiful and attractive. Impressions of dry flowers on all the pages of the book can also make books look beautiful and attractive.

22.11.4 Cards

Cards are a wonderful and the most beautiful way to express one's feelings. Dried florals are used for decorating cards with a wild imagination. One can make various designs using dried flowers on the cards and the cards can be personalized by even writing your name using dried flowers. Cards made using dried flowers can add warmth to the cards.

22.11.4.1 Making Cards with Dried Flowers

- The cards are cut to any shape or size.
- Select some dried flowers and arrange them on the front of the card.
- The desired picture is obtained by gluing flowers and leaves to the card.
- The card when laminated protects the dried material and adds to it's beauty.

22.12 Photo Frames

Handicrafts made by the artisans and craftsmen in India are in great demand worldwide. Handmade dried flowers photo frame is a major industry in India. Different types of cloths are used to make dried flowers photo frames.

Different colors and kinds of dried flowers are used to beautify the photo frames. They look elegant and bring about grace to any occasion and setting. They create charisma out of dried summer hues and beautiful designs can be made using them. They can be made in any desirable shapes. They serve as the best gift one can give on weddings, anniversaries and birthdays. Any handmade product adds a personal touch to the same.

22.12.1 Making a Photo Frame Using Dried Flowers

- A piece of cardboard is taken and cut into the desired shape and size. It is covered with calligraphy paper or any other handmade paper.
- Another piece of the same size and shape is also cut.
- On the other piece of cardboard a small piece is carved in the center of the cardboard to fit in the photograph. The remaining portion is covered with a calligraphy paper.

- Thick cellophane or any transparent sheet is put on the back of the carved cardboard.
- Both the cardboard pieces are stuck together leaving one end open.
- On the front of the photo frame dried flowers or leaves are arranged and glue according to one's taste.
- A photograph can be mounted on this frame now.

22.13 Candle Holders

Using dried flowers for making candle holders is one of the artistic and sophisticated idea that is popular these days.

22.13.1 Making Candle Holders with Dried Flowers

- To make candle holders with dried flowers a wire is rolled it into a tube.
- The tube is bent so that the ends meet and joined together to make a ring.
- A spot for the candles is created by cutting short pieces of pipe.
- The pipe is pushed between the wire mesh and the candles set in place.
- The holder is dressed up with dried flowers and foliage.
- Ribbons can be used to add to decorate the candle.

22.14 Carry Bags

Use of dried flowers adds style, uniqueness and unmatchable quality to the carry bags. Carry bags are used by all ages and their various designs and styles are suited to all tastes. They can be made from jute, cotton, canvas and many other materials. Elegant pieces of dried flower bags are available today in the worldwide market. These are offered in a variety of colors and features beautiful designs made by dried flowers. They are great to store and protect various items and also provide an additional aura of beauty. Carry bags made with dried flowers is an art that stretches to antiquity and eternity.

22.15 Wall Quilt

A quilt is a fabric sandwich made of three layers, a top, a filling, and a backing. These layers are held together by what is called quilting, stitching that goes through all layers. A wall quilt is a quilt made with smaller dimensions and meant for hanging on a wall. Wall quilts can contain specialty fabrics and embellishments that are not meant to be washed or undergo strain or wear. Wall quilts made of dried flowers introduce a little of summer hues in the home and add zest to the living room. They are available with many botanical prints and its fun to make these quilts.

22.15.1 Making a Wall Quilt

- A frame is used for this.
- The backboard of the frame is removed.
- The fabric is stretched over the front of the board.
- The edges of the fabric is folded to the back of the board and glued with a glue gun.
- A desirable square of the cardstock is cut.
- Pressed flowers and leaves are arranged on the squares of the cardstock.
- The dried flowers and leaves are carefully glued to the cardstock.
- The flower squares are placed on the fabric-covered board and each square arranged so as to be evenly spaced.
- Each square is glued to the fabric.
- The frame is replaced onto the backboard.

22.16 Topiary

Topiary is the skilled art of clipping or trimming foliage or flower arrangements so that they take on take on the shapes of animals, lettering, numbers, or various but precise geometric forms. The early colonists brought the ancient art of topiary to this country. One can carry on the tradition with this beautiful created topiary. They can have a variety of dried flowers arranged. Fanciful topiaries always seem to fascinate and spark the imagination of delightful gardeners and plant lovers.

22.16.1 Making a Topiary

- Four styrofoam balls are used for this. One of them is sliced with a serrated knife to one third of the way down.
- A stick is pushed into the center of the ball so it stands straight. A little glue around the stick is applied to hold it in place.
- The second ball is sprayed green and small holes made in it.
- The boxwood branches are dipped into hot glue and placed in the holes of the ball continuously till the whole ball is covered.
- A variety of dried flowers is added, balancing colors around the ball. Each flower is dipped in hot glue and placed among the boxwood branches.
- The boxwood and flower is pushed into covered ball onto the stick in the pot.
- The same is repeated with the rest of two balls.
- Small pieces of moss is added to hide the bottom ball and create a finished look.

22.17 Jute Bags

Jute is consistently available in new and exciting form to create trendy bags. Jute bags designed using dry flowers look very elegant and can be made with innovative designs. Jute is considered to be a gift bestowed by nature. It is seen as the most ecological fabrics that can be put into unlimited usage. Jute and dry flowers décor together form a completely eco-friendly bag. Using dried flowers, leaves and foliage a number of designs and patterns are made on the jute bag. They can also be customized by a special message written by dried leaves.

22.18 Dried Floral Tree

Small trees made with real branches for the trunk and a cone-shaped piece of dry foam for the tree is easy and charming. This requires some plaster of paris (not pre-mixed), a container, some moss and floral pins to secure the moss and small and dried flora for tree decorations.

1. The container should be first lined with slivers of floral foam. The plaster of paris poured into it expands as it sets. The foam acts as a cushion and keeps the pot from cracking.
2. Enough plaster of Paris and water is mixed to fill the pot two thirds full. The slivers of foam should stay in place as the mixture is poured into the pot.
3. The trunk of the tree is inserted, making sure it rests on the bottom of the pot. The pot is turned and adjusted until it looks straight from all angles. Plaster is added to 1" of the top of the pot.
4. When the plaster is dry, the cone is pushed onto the trunk, and covered with moss. Floral pins are used to hold the moss in place.
5. The tree is decorated by hot gluing dried flowers and leaves to the moss. The top of the plaster is covered with glue and add moss and a few flowers.

22.19 Other Gifts

Dried flowers can be customized and serve as a perfect gift. An accent piece of dried flower arrangement serves as a wonderful gift. Shirts imprinted with dried flowers look amazing. A number of gift items can be made using dried flowers. Some of the items include tall winter birdhouse, doll, bags, welcome wreath, candle jar, and trays.

Gift Items

- Fanciful topiaries always seem to fascinate and spark the imagination of delightful gardeners and plant lovers.
- Beautifully made notebooks with dried flowers handmade paper to record the memorable moments.

- Wall quilts made of dried flowers introduce a little of summer hues in the home and add zest to the living room.
- Handmade photo frames made using dried flowers are the best gifts for weddings, anniversaries and birthdays.
- Candleholders add ambiance to homes and occasions like birthdays and Christmas.
- Elegant pieces of dried flower bags are available as a perfect gift.

Summary

Whether used in bouquets, in potpourri, on wreaths, dried flowers inspire us to create beauty and elegance in our homes. Flowers and foliage are dried to capture the essence and beauty of flowers in winters. Dried flowers together with decorative grasses, berry-laden branches of shrubs, and foliage, make great decorations. Using dried flowers one can make swags, baskets, bouquets, wreaths, and garlands to enhance the décor of your home and offices. Floral wires are used to bind a group of twigs and branches together. To make it more beautiful and attractive pinecones, waxed leaves, bells, bows, spray painted wooden toys are added. Sprays are also used to add fragrance to it. A selection of interior design baskets and other whimsical containers are used to enhance any interior design scheme. These arrangements form the best décor accents and gifts. They add a touch of elegance, enhanced beauty and cheerfulness to any office or home. They form a unique gift that adds an undeniable charm and brings a smile to just about any occasion. Moss, twigs, wheat and grasses also enhance the dried flower arrangement. Adding a few silk flowers, ribbons or bows to the arrangement ensures a more delicate and sophisticated look. They help create a warm welcoming feeling in one's home.

Questions and Answers

1. Write about different types of dry flower decorations and the method of preparing them.

Chapter-23

Packing and Storage

23.1 Dry Flower Packing

Careful packing is one of the most important aspects of any product. Dry flower packing is the simplest way to capture the essence of summer flowers till long. Dry flowers should be packed nicely to avoid any damage. They should be packed alone in appropriate sized carton loosely cushioned with paper.They should be packed in a manner to keep them away from direct sunlight leaving some room for air. Packing material should be stable and firm to hold the dry flowers well without any damage during storage or transportation. Dry flowers can be packed in a variety of ways: in poly bags, window box, cardboard box, or in a sachet.

Dry Flower Packing can be in

- Poly Bag
- Sachet
- Window Box
- cardboard Box

23.2 Poly Bag

A poly bag is a bag that is made of polyethylene, which is a flexible, water resistant plastic that is ideal for making merchandise bags. Poly bag is a traditional way to capture the fragrance of summer hues. They are widely used in packaging of dry flowers for a longer and safer storage.

23.2.1 Types of Poly Bags

- Low Density Polyethylene- It is the material used in most poly bags. It has greater tensile strength, clarity, flexibility and water-resistant properties.
- High Density Polyethylene- It is made from a thinner gauge material to lower freight costs and storage requirements. It has great strength and is water-resistant.

23.2.2 Packing of Dry Flowers

Dry flowers are safely packed in poly bags. Dry flowers are tied together with a rope or a thread by their stems. They are gently placed in the poly bags for storage or shipping. The poly bags are then closed with a tape or are sealed at the open end. For packing dry flowers generally low-density polyethylene bags are used.

23.3 Window Box

Window boxes are gardens in a box, manageable and enchanting. They can be planned to co-ordinate with the home's color scheme, accent a scene, celebrate an occasion, provide decoration for the kitchen or entrance. Various types and colors of dry flowers are beautifully arranged in the window boxes and can be stored for longer. Window boxes decorate windows, railings, fire escapes and patios. Houses, apartments and even sheds can be enhanced with the addition of a flower filled window box, spilling over with dry grasses and foliage. The window box adds color to the room. They add zest to your home and can firmly hold dry flowers and protect them.

Window boxes are made of highly durable fiberglass. The fiberglass is strong and sturdy to take the weight of dry flowers and added marbles or pebbles. They won't rot and are backed with a lifetime guarantee. Fiberglass window boxes look better and more balanced when they are little wider than the actual window and the style of the window box should match the architectural integrity of the home.

23.4 Cardboard Box

Cardboard is a paper product made of unbleached craft paper, with two heavy outer layers and inner layer to provide strength. Boxes are made of cardboard for packing of dry flowers. Cardboard boxes are the most versatile of all packing materials for dry flowers. They can store dry flowers well for a longer period and are easy to ship long distances. They protect dry flowers from all possible hazards.

23.4.1 Method of Packing

Small bunches of dry flowers are wrapped in tissue paper or newspaper and tied with a little string or raffia. The material is stored in a cardboard box, stacking them very carefully rotating the floral ends. The box is sealed or a tape applied to it. These cardboard boxes can also be made tamper proof and logos or symbols can be printed on the box for the identification of different types of dry flowers.

23.5 Storing and Preserving Dry Flowers

Flowers have lots of moisture in them. When they are put into a drying mixture, the moisture draws and the flowers dries out. Drying of flowers keeps these beautiful flowers alive and they don't get moldy. But drying flowers emphasize imperfections. When drying a flower, many flowers lose their color or texture and at times colors fade. Flowers that have started to wilt or die are not suitable for drying. Storing dry flower products properly is very important. Since a lot of effort time has been spent to dry them, it is necessary to protect them from all possible hazards. Floral boxes are also available that are best for storing dried flowers. They can even be well packed in poly bags, window box, or in a sachet.

They are stored in airtight container to prevent it from moisture and sunlight. They are placed in a dark, dry place and a well-ventilated place. Preserving flowers is a wonderful way to recycle natural materials and enjoy them year around by arranging them in vases, creating wreaths, candle holders and other home decorations. Flowers are not always available year around and preserving them allows enjoying a larger range of flowers in and out of season. To preserve flowers a drop of bleach is added to water. Some flowers are then bleached to dry them and color is injected into them at a later stage.

23.5.1 Tips on Preserving Dry Flowers

- Dried flowers should be kept out of direct sunlight and a path where people don't brush against them.
- Don't place them in the path of forced air heat. The hot, dry air just dries them out too much.
- Dried flowers can last up to several years if cared properly. A clear acrylic sealer to can be utilized to preserve them.
- They should be kept fresh and dust free using simple misting with fresh water.
- Loofa sponges should not be harvested until the vine dies back and the fruit completely dies.
- Rose petals or other fragment flowers should not be disposed off. Instead they should be dried.
- Hydrangea flowers with woody stems dry beautifully when allowed to stand upright in a vase with little water. The flowers soak up all of the water and remain in the vase until they are totally dried.
- Strawflowers intended for dried flower arrangements should be cut off when the blooms are only half-open. Small bundles of the flowers are tied together and hanged upside down in a well-ventilated place to dry.
- During the winter, flowers should be dried over a heat vent to retain a brighter color.
- Flowers to be dried are dyed with absorption dyes before pressing or drying to compensate for fading. Bleaching agents are essential to remove contamination from the blooms.
- Ferns and foliage that are prone to fading are placed into floral absorption dye before drying.
- Dye is added to the warm water and freshly cut stem is put in the water and allowed to sit a few hours.
- To reduce brittleness in the pressed fern or foliage, a little glycerin or fabric softener is added to the solution.

Summary

Careful packing is one of the most important aspects of any product. Dry flower packing is the simplest way to capture the essence of summer flowers till long. Dry flowers should be packed nicely to avoid any damage. They should be packed alone in appropriate sized carton loosely cushioned with paper.They should be packed in a manner to keep them away from direct sunlight leaving some room for air. Packing material should be stable and firm to hold the dry flowers well without any damage during storage or transportation. Dry flowers can be packed in a variety of ways: in poly bags, window box, cardboard box, or in a sachet. Storing them in box or stacking them very carefully is important. Floral boxes are also available that are best for storing dried flowers. They can even be well packed in poly bags, window box, or in a sachet.

Question and Answers

1. Write a short note on storing and packing of dry flowers.

Chapter-24

Drift Wood

Driftwood is wood that has been washed onto a shore or beach of a sea or river by the action of winds, tides, waves or man. It is a form of marine debris or tide wrack.

In some waterfront areas, driftwood is a major nuisance. However, the driftwood provides shelter and food for birds, fish and other aquatic species as it floats in the ocean. Gribbles, shipworm sand bacteria decompose the wood and gradually turn it into nutrients that are reintroduced to the food web. Sometimes, the partially decomposed wood washes ashore, where it also shelters birds, plants, and other species. Driftwood can become the foundation for sand dunes.

Most driftwood is the remains of trees, in whole or part, that have been washed into the ocean, due to flooding, high winds, or other natural occurrences, or as the result of logging. Other sources include the remains of man-made wooden objects, including buildings and their contents washed into the sea during storms, wooden objects discarded into the water from shore (flotsam), dropped dunnage or lost cargo from ships (jetsam), and the remains of shipwrecked wooden ships and boats. Erosion and wave action may make it difficult or impossible to determine the origin of a particular piece of driftwood.

Burning driftwood can produce polychlorinated dibenzodioxins (PCDDs), also known as the misnomer "dioxins", which are carcinogenic. For this reason burning driftwood is not recommended. The formation of PCDDs is well documented when organic compounds are combusted in the presence of chlorine, which is present in driftwood as a result of soaking in sea water.

Driftwood can be used as part of decorative furniture or other art forms, and is a popular element in the scenery of fish tanks.

Sculptures Using driftwood

Chapter-25

Review of Research in Dry Flower Technology

Preserving plant materials in a dried form is not a new idea. It has been considered as an art for hundreds of years. Fragrant dried herbs were placed in Egyptian pyramids. Dry flower arrangements have been popular in Europe for several centuries. During 1700 A.D, the early American settlers used dry flowers to brighten their homes. Any plant material can be preserved for use as dry flowers (Al Pertuit, 2004). Though enormous work is being done in this field, most of the literature is published by horticulture amateurs. Scientific work in this area is less. However, a review of the research work done in dry flower technology is discussed in detail.

25.1 Drying Methods

25.1.1 Air Drying

Bryan (1992) described air drying as the easiest and most effective method for astilbe, delphinium, gypsophila, helichrysum, mimosa, roses, statice, grasses, seed pods and leaves. Blacklock (1993) reported that roses dried in seven days and other smaller flowers took less time. Sell and Aakre (1993) described this method. Flowers and foliage are cut uniformly. Then they are hung upside down in a dark, dry aerated room. This method is suitable for eranthemum, globe amaranth, golden rod, hydrangea, salvia and statice. Bhutani (1995) observed that flowers with a crisp texture like *Helichrysum* sp., *Helipterum* sp. and *Limonium* sp. could be dried upright. Anon., (1996) revealed that gypsophila and statice could be dried by keeping erect in a jar. Astilbe, cockscomb, globe amaranth, golden rod, marigold and yarrow grasses could be dried by hanging upside down. Seaberg (1996) stated that only perfect rose flowers harvested on dry days dried well. The flower bunches tied with rubber bands after removing the lower leaves were hung upside down in a dry dark room. It took 5-10 days to dry. Mckenzie (1996) standardized techniques of drying for many species.

A study was conducted with loose bunches of stems of flowers tied with a string, rubber band or twist tie. This was hung upside down in a warm dry place where air circulated freely. Results showed that time to dry varied from a few days to 2 to 3 weeks depending upon the flowers (Anon., 1997a). Evelyn (1997) attempted the method of air drying in dusty rooms. For that, flower bunches were placed in a paper bag which was perforated all around for air circulation and tied around stems. The seed heads dried in two weeks depending on the weather. Macphail (1997) evaluated 30 cultivars of 16 plant species and found that acrolinum, bells of Ireland, gomphrena, gypsophila, helichrysum and statice, dried well by hang dry method in a dark dry room. Rothenberger (1997) reported on the harvesting techniques and drying methods for dry flowers. He suggested that good quality flowers at slightly immature stage should be selected. Foliage should be removed. Stewart (1997) observed that simple air drying of coloured roses caused shriveling and colour darkening. Bright red roses usually become the colour of dried blood. White roses usually become a yellow parchment colour. Thomler (1997) suggested that air drying of upright growing flowers dried best upside down. Ornamental grasses, cat tails and branches dried well by upright method. The leaves of some trees and shrubs like oak cuttings should be placed flat in a shallow box lined with sheets of blotting paper. Soft flat topped flower heads such as Queen Ann's lace, dried best when their stems were inserted in wire mesh. Yockley (1997) listed annuals like amaranthus, dusty miller, nigella and statice and perennials like artemesia, baby's breath, chinese lantern, lamb's ear, lady's mantle, oregano and yarrow. They were quite suitable for natural drying.

Rubber bands were used to make small bunches which would pull tighter as stems shrink during drying. Bunches should be hung upside down in a dark area with good air circulation. It would take 2

-3 weeks to dry completely (Anon., 1998). John Van Dyk (1998) reported that flowers can be dried within 2-3 weeks. Kiesar (1998) found shade drying to be cost effective for gomphrena flowers. Verey (1998) suggested drying by hanging upside down for achillea, amaranthus, astilbe, calendula, dahlia, delphinium, eranthemum, gypsophila, helichrysum, physalis and salvia. Drying upright was beneficial for gypsophila, hydrangea, lunaria and rose. Drying flat was reported for artichokes, deciduous leaves, maize, proteas and thistles.

Datta (1999) reported air drying as the most simple and cheap method of dehydration. Drying upright was beneficial for gypsophila, hydrangea, lunaria and rose. Drying flat was reported for artichokes, deciduous leaves, maize, proteas and thistles. Alleman (1999) recommended this method for althea pods, oats, pampas, grass statice, straw flowers and yucca pods. The drying time varied according to the moisture content of the dried material and weather condition. Bull (1999) found that the drying process took from 10 to 20 days, depending on the plant, for shade drying. The test for dryness was commonly done by dissecting one or two flowers and thorough dryness was ensured. Weak flowers are wired before drying. Jasmeet (1999) found that the most common method of drying was air drying, where the flowers and grasses were tied loosely with thin wire and are hung upside down. However, this method was dependent on the prevailing climatic conditions. Rengasamy et al., (1999) reported that flowers with crisp texture dried easily by hanging upside down or placing stems erect. This method was cheap. However, petal shrinkage was the disadvantage. Helichrysum, statice and flowers with less petals and papery texture dry well.

Venugopal and Patil (2000) studied the effect of drying methods and temperatures on colour intensity of everlasting flowers. Drying at room temperature under shade and oven drying at 50°C retained the colour intensity for 150 to 180 days. Fresco, 2002 reported that air drying is commonly referred to as the "hang and dry" method. It is the oldest and easiest drying technique. There is no special equipment needed. However, it is one of the longest drying methods. It usually takes three to four weeks for the flowers to dry completely. Air dried products tend to loose their original colour faster. Irene Vethamoni and Nanthakumar (2002) standardized drying techniques in ten crops *viz.*, achillea, artemesia, buddlea, daisy, helichrysum, hydrangea, lupinus, marigold, salvia and solidago. Sun and shade drying was good for lupinus, salvia and solidago. Doreen (2003) reported that peony buds and blossoms retain good color and shape when dried. The double types are used for Victorian style bouquets. The single and semi double types are used for contemporary arrangements. Both buds and blooms can be dried by hanging upside down or by the surround and cover method. For commercial use, bunching and grading is done in the field. Flowers are dried immediately. Foliage is stripped from the bottom one third of the stem. Flower heads are bunched together. Stems are cut to uniform length. Two rubber bands are tied for each bunch. One at the cut end and one at the flower end. The flower bunches are hanged to dry in a dark room at 50°F or higher with 50 to 60 percent relative humidity (Doreen, 2003).

Irene Vethamoni et al., (2003) conducted trials on identification and standardization of plant species suitable for dry flowers at Kodaikanal. Among the foliage studied, shade dried *Cyperus* sp. and *Thuja* sp. had the best retention of green colour. Similarly among the flowers studied, helichrysum, lupin, salvia and solidago were the most suited for drying irrespective of the method of drying followed. Singh and Dhaduk (2005) recommended hang drying under shade for globe amaranth. According to IIHR research findings, air dried *Celosia* sp. gave good quality dry flowers (Anon., 2006a).

25.1.2 Silica gel and borax Drying

Rebecca *et al.,* (1995) experimented on drying rose buds and carnations by air drying and embedding in sand. Drying by embedding in sand gave better results than air drying. Samant and Dash, (1998) studied the effect of desiccants on drying of roses. Flowers dried perfectly after two weeks of embedding with silica gel and with silica gel and sand (1:1). Chimonidou - Pavlidou (2000) reported that species like *Achillea* sp., *Delphinium* sp., *Helichrysum* sp., *Helipterum* sp. and *Limonium* sp. could be dried in a dark ventilated room at 30^{0}C for the first week. For the following 2-3 weeks drying at 20^{0}C maintained colour and quality. Chitra (2001) recommended silica gel drying for small and large flowered types of *Plumeria alba,* a non-traditional raw material. Gill et al., (2002) evaluated various methods of drying flowers and foliage and concluded that embedding in silica gel was the best for carnation, ferns, helichrysum, rose and silver oak.

Doreen, (2003) reported that single and semi double peonies should be dried using the surround and cover method. This is the best method to preserve the shape. Partly open double flowers are also used for drying. The flowers dry most efficiently in shallow containers. So, most of the stem is removed. Drying substances such as white cornmeal, sand, borax, kitty litter, and silica gel are used. Substances that will harm the flower or be difficult to remove should be avoided. The stems are removed. Flowers are placed on a 1.2 to 3.4 inch layer of drying substance in a container. The container should be two to three inches deep. The drying substance is carefully poured over, around, and through the petals to cover the flowers.

Keisar *et al.*, (2003) revealed that silica gel drying of french marigold and zinnia gave the best aesthetic qualities. Singh *et al.*, (2003) studied the effect of dehydration on postharvest life and quality of zinnia flowers. Results revealed that drying was much faster with silica gel and borax as compared to sand. Silica gel of size 60 to120 mesh, was the best absorbent for removing moisture from flower and foliage followed by boric acid. Raj and Gupta (2005) standardized the dehydration technology for ten ornamental herbaceous plants and twenty eight tree species from Outer Himalayas. They concluded that silica gel drying was the best method followed by boric acid drying.

Singh and Dhaduk (2005) studied the effect of various drying techniques on flowers like bauhinia petals and whole flower, gomphrena, ixora, mussaenda, plumeria and sunflower. Embedding in silica gel was suitable for maintaining three dimensional view of flowers of bauhinia, plumeria and sunflower. Saffena et al., (2006) studied the response of drying roses in hot air oven and concluded that silica gel drying in hot air oven at 40^0C resulted in the best quality flowers.

25.1.3 Microwave and Hot Air Oven Drying

Gill *et al.* (2002) reported that asparagus leaves dried best in microwave oven for 1hour 30 minutes. Irene Vethamoni and Nanthakumar (2002) standardized drying techniques for the following plants *viz.*, achillea, artemesia, buddlea, daisy, helichrysum, hydrangea, lupinus, marigold, salvia, and solidago. According to them, oven drying helichrysum gave best results. Moisture loss was the highest in oven-dry flowers. Dahiya *et al.* (2003) reported that flowers of annual chrysanthemum dried in hot air oven by embedding in silica gel at 50^0C for 48 hours resulted in the best quality flowers. Fully bloomed flowers dried by microwave drying for 6 to 7 minutes recorded highest scores for aesthetic and visual qualities (Geetha *et al.*, 2003). Tight buds of yellow helichrysum flowers recorded maximum sensory scores of 5 when embedded in sand and dried in hot air oven (Sangama, 2004).

Raj and Gupta (2005) standardized the dehydration technology for ten ornamental herbaceous plants from Outer Himalayas. In most cases, drying in solar drier was at par with drying in hot air oven at 45°C or microwave oven (450 watt). Solar drying was also found to be economically feasible. Singh and Dhaduk (2005) concluded that press drying in shade or oven was suitable for maintaining two-dimensional view of petals of *Bauhinia purpurea, Ixora, Mussaenda* and *Myenia erecta*. Bhalla *et al.*, (2006) reported that chrysanthemum should be first embedded in silica gel. Then it should be dried in hot air oven at 60°C for 24 h. In microwave oven the drying time is 90 seconds. Studies were conducted on the technology for dry flower production of gerbera (*Gerbera jamesonii*).

25.1.4 Glycerin Drying

In this method moisture in a flower or foliage is replaced with glycerin and water. The flower is preserved and not dried. Kawasaki (1975) reported that glycerin results in longer drying of the leaf. The process of glycerin drying is as follows. The bark is removed. The cut end is smashed using a hammer. This portion of the stem is immersed in the glycerin and water solution. The leaf turns brown with glossy appearance after a month. Conder (1979) tried glycerin method of drying. He reported that dried materials (whole bunches or single leaves) retained their natural shape and flexibility. They lasted indefinitely and could be dusted or even wiped with a damp cloth without risk owing to the leathery texture of leaves. Dana (1983) reported that many types of foliage have been successfully preserved to maintain softness and flexibility. This is done by immersing leaves or placing stems in a 33 per cent glycerol solution.

Wilkins and Desborough (1986) tried using glycerine, clove oil, ethylene glycol alone or in combination along with dimethyl sulfoxide before cryodrying. They compared this with untreated cryodry flowers. Flowers cryodried without pretreatment were aesthetically acceptable. However, Paparozzi and Mc Callister (1988) established that freshly cut statice stems, should be preserved by soaking in 1:2 or 1:3 glycerol: water solution for 48 hr followed by microwave drying for 1 min at medium high temperatures (34°C). Semant et al., (1993) reported that one part of glycerin mixed with two parts of hot water was suitable for twigs of 26 plant species at room temperature. Full absorption was recorded between 15 and 30 days after treatment. This depended on the plant species and its age. Verey (1998) recommended the use of 1:2 glycerol: warm water dipping to preserve eucalyptus, hollyhock, hydrangea etc. Addition of a few drops of vegetable oil intensified the colour of immersed stems (Miller,1999). Padvathamma (1999) concluded that 33 per cent glycerol solution was the best to dry statice flowers.

Pre-treatment with 1:5 ratio of glycerol and water in rose var. Noblesse improved the colour, size and flexibility in dry flowers. Chrysanthemum var. yellow double resulted in smooth, bright and large size dry flowers when dried in hot air oven with sand embedded technique after pre-treatment with glycerol and water mixed in the ratio of 1:5. Similarly, air dried helichrysum pre-treated with 1:4 ratio glycerol was the best (Anon., 2006a).

25.1.5 Freeze Drying

Freeze drying was first used by the Peruvian Incas of the Andes, who stored their food on the heights of the Andes. The extreme cold temperatures at that height froze the vegetables, and the water inside slowly vapourized under the low air pressure. During World War II, freeze drying was commercially used to preserve blood plasma. Since then freeze dried products are used commercially and also for space programs. The preservation of flowers by this method was evolved in the 1990's. This is primarily because the flowers needed treatments before and after the process to avoid shattering and loss of colour (Tom Harris, 1999).

Wilkins and Desborough (1986) conducted experiments on cryo drying of *Dianthus caryophyllus*. Several cultivars of carnation were successfully cryo dried and remained natural in appearance after being placed at -80°C for 12 hours and then placed in a freeze drier for 7 days. Anon. (1997b) described the method of freeze drying of roses and reported that such flowers lasted 4-5 years or more. However, they should never be placed in humid areas. Anon. (1997c) reported freeze drying as the most technologically advanced way of drying flowers. Flowers like camellias, peonies and hydrangeas can be freeze dried resulting in darker colours that resist fading.

Thomas (1999) has reported that there has been limited success in freeze drying of floral products; only the flower portion of the plant appears to respond to the drying process, and there is often a slight change in colour. eg. red roses tend to take on a purple tint. The stems of the flowers tend to assume a brown tint and the leaves become brittle and curled. In drying of plants, the nature of the plant plays an important role eg., cacti is not suitable for freeze drying. Hence, to achieve a more natural product with a relatively longer shelf life, various pretreatments might be necessary.

Chen *et al.,* (2000) studied the effects of different freeze drying processes on the moisture content, colour and physical strength of roses and carnations. Vacuum drying temperatures had more effect on the flowers than freezing time. Lower vacuum drying temperatures resulted in flowers with colour closer to fresh and control flowers. Higher temperatures resulted in low moisture contents and stronger, stiffer petals but more changes in colour. Freeze drying works well for peonies, but requires expensive equipment, Doreen (2003).

25.1.6 Press Drying

Trinklein (2006) reported that pressed flowers are especially suitable for flower pictures, greeting cards etc. Flowers that are in prime condition are used for pressing. Also, flowers with different stages of development up to full maturity are used for more variety in design. Plants with fleshy stems and leaves, flowers with very thin petals and wilted materials are not pressed. Flowers that are flat, such as pansies, are best for press drying. The faster flowers dry, the better they retain color. On the other hand, flowers can't be exposed to excessively high temperatures; although they may dry quickly, they will turn brown.

Pressing requires sandwiching flowers and foliage between layers of an absorbent material. This should be clean and hold the flowers firmly and flat during the drying process. Porous materials that allow some air movement are also beneficial. Flowers are generally placed between sheets of a non glossy type of paper. Newspapers, old telephone directories or catalogs are suitable. Absorbent facial tissues placed on the pages aid rapid moisture absorption. Tissues should be removed and flowers or foliage replaced between fresh, dry tissues and papers at the end of the first week. After the flowers and tissues have been placed in the folded newspapers or books, they are stacked several layers deep. Boards are placed beneath and on top of the stack. The stack is placed in a warm, dry place with a heavy weight on top.

Another satisfactory system suitable for drying flowers uses a combination of cardboard, newspaper and desk-sized blotter pads.

Corrugated cardboard is cut into sheets slightly larger than the sheets of folded newspaper. Flowers are positioned on one side of the opened newspaper. Then, the newspaper is closed and a sheet of blotter paper placed on either side. After all the flowers have been placed between the newspapers, blotter paper and cardboard, the layers are stacked and tied or taped together. They should then be placed in a warm, well-ventilated place. A heavy weight is placed over this. Special presses can be purchased or constructed for drying large quantities of materials.

25.2 Bleaching

As early as in 1967, Robson claimed that bleaching occurs due to chemical modification of the coloured compounds within the tissues. To reduce the processing costs, he suggested multistage bleaching. In this method, both oxidative and reductive bleaching chemicals are used. In his opinion, the use of peroxide at room temperature gave good results. Gulrajani and Sukumar (1985) ascertained chlorite bleaching to be the best chemical method for whitening of delicate leaf tissues with the least physical damage. Venkataraj and Gulrajani (1989) reported the use of foamed sodium chlorite (80 per cent) composition for bleaching polyester or cotton blended fabrics. Mirunalini (1999) studied various drying techniques for celosia, rose and chrysanthemum at four stages of harvest and subjected them to various bleaching and dyeing experiments. They reported that the half and full bloom stage were ideal for drying, bleaching and dyeing of the three crops. Bleaching in 10per cent sodium chlorite for nine hours treatment was best. Yogita Chari (2000) conducted studies to standardize bleaching and dyeing technology of hybrid tea roses and *Aerva* sp. respectively. They concluded that sodium chlorite was the most ideal chemical for bleaching roses. Of the various bleaching mediums used, distilled water was the best.

Chitra (2001) conducted studies to evolve appropriate drying and bleaching techniques for small and large flowered types of *Plumeria alba,* a non-traditional raw material. Of the three bleaching methods tested, sodium chlorite 30 per cent was found to be an effective bleaching agent and also faster than other bleaching agents. Calcium hypochlorite and hydrogen peroxide resulted in poor quality flowers which were unsuitable for drying. Sujatha et al., (2001) established that hydrogen peroxide was the best bleach at room temperature and caused the least damage to cell tissues. Keisar et al., (2002) recorded that sodium hypochlorite was the best chemical to bleach pink globe amaranth, french marigold and multi coloured zinnia. Bleaching and dyeing treatments were implemented on celosia for value addition by Priyesh *et al.* (2003). Bleaching with 30 per cent hydrogen peroxide scored the highest values in terms of bleaching qualities.

25.3 Dyeing

William and Segal (1972) reported that the basic dyes were the first of the synthetic type to be made out of coal tar derivatives. Although basic dyes produce brilliant colours, they have poor fastness to light and washing. The direct dyes do not produce very brilliant colours but have better fastness to light and washing. Colour fastness of acid dyes is generally increased by adding metallic salts to the dye stuff. The sulphur dyes have excellent resistance to washing but not to sunlight. Azoic dyes or the naphthols are primarily used for bright red shades. The vat dyes are known for their fastness to both washing and sunlight. They are made from indigo anthroquinone and carbazol. The reactive dyes can be used such that the whole spectrum of colour can be put onto cloth using just one class of dyes. The other textile dyes include acetate dyes, alizaric dyes, aniline block, chrome dyes, soluble vats, neutral dyes, disperse dyes and indigo.

Mark *et al.* (1979) reported the acid dyes to be water soluble anionic dyes, while the vat dyes are water insoluble dyes. Anilker (1981) reported acute toxicity among the basic dyes and diazo direct dyes. Westland (1992) furnished various methodologies of dyeing the dry flowers. The flowers of gypsophila and hydrangea can be made to absorb the dye from the dye solution mixed with water. Dip dyeing and spraying may be recommended to colour the seeds and pods. Natural dyes and commercial dyes may also be used.

25.3.1 Chemical Dyes

Mirunalini (1999) reported dyeing by immersion in vat group of dyes for ten minutes at 0.2 per cent concentration. Yogita Chari (2000) conducted studies to standardize bleaching and dyeing technology of hybrid tea roses and *Aerva* sp. respectively. Of the various dyes used, the basic group of dyes was the best at 3 per cent and 0.3 percent for bleached roses and *Aerva* flowers respectively. Among the various dyes used, vat group was superior followed by direct and acid colours. These groups had low level of colour fading on storage. The time taken for drying after dyeing was similar in all the groups (Keisar *et al.*, 2002). In dyeing experiments implemented on celosia by Priyesh *et al.* (2003), vat groups of dyes scored the highest values in terms of visual aesthetic qualities. Sangama (2003) studied dehydration of *Gomphrena globosa* flowers and reported that all stages and all methods of drying were acceptable. However, shade dry flowers were better over sun and oven dry flowers.

25.3.2 Natural Dyes

Khanna and Chandra, (1996) reported that an upsurge in interest in natural dyes has been manifested as reconstruction of ancient and traditional dyeing technology, growing use of natural dyes in craft work and replacement of synthetic dyes for food safety. Natural dyes have become a part of human life since time immemorial. These dyes obtained from natural sources are non-pollutant, non - allergic, having rich shades (Gupta *et al.*, 1997). Due to carcinogenic nature of synthetic dyes and their intermediaries there is enough scope to revive the age old art of natural dyeing. The dye potential of natural dye extracted from marigold was evaluated by dyeing the dry flower of *Lagurus ovatus* under different dyeing conditions. The chromaticity values and colour difference of dyed flowers were higher than those of control. The surface colour was not affected by soaking time. Flower quality was maintained even when the flowers were soaked for only 5 minutes soaking. Surface colour was the best when dyeing was done at 61^0C to 70^0C (Song *et al.*, 2004).

25.4 Dry Flower Products

25.4.1 Pot pourri

Pot pourri is a special type of dry flower arrangement which is made as a decorative piece and a room freshener. Pot pourri is usually a mixture of dried, sweet scented plant parts including flowers, leaves, seeds, stems and roots. The materials used for this purpose should be resistant to mould, non toxic and able to withstand mechanical damage. The basis of pot pourri is the aromatic oils found within the plant (Geetha *et al.*, 2004).

Boonsong, (2008) conducted studies on the improvement of aroma in Thai Pot pourri and dry flowers using starch film as an aroma fixative. The results of the study indicated that the starch mixed with essential oils of cloves constituted the best aroma fixative for potpourri and dry flower mixture.

25.5 Flower Arrangements

Work on use of dehydrated material as flower arrangement was carried out by Kher and Bhutani in 1979 at NBRI, Lucknow. Dried flower arrangements are prepared on a base. Since they have to be preserved from dust, humidity and wind, they are covered with a slightly oversized transparent glass or plastic container. The container is placed upside down on the arrangement and attached to the base using an appropriate adhesive. It is then kept in the hot air oven at 45^{0}C for about half an hour. Enamel paint is used to paint the bottom half of the arrangement.

25.6 Dry Flower Storage

Trinklein (2006) explained the method of storing dried material. Dried materials in storage may occasionally be attacked by one or more household insects such as museum beetles, silver fish, cockroaches etc. As they chew on the soft tissue of the plant centres, flowers may shatter and fall apart. They do not necessarily occur on the flowers as they come from the garden, but move into the boxes in the home during storage. To overcome this problem, boxes should be checked occasionally and if insects are present, the infected materials should be destroyed. Tightly sealed containers prevent infestation. If the pests are present, the container should be thoroughly cleaned before using it again. Naphthalene flakes may repel insects, or some general insecticides may help control them, but once an area is infested, complete eradication is difficult. Dry flowers, should be preferably replaced yearly, but with good care they often last longer. Even the best dry flowers gradually fade and should be discarded when they no longer produce the desired effect. Flowers that tend to fade may be lightly tinted with aerosol paints or dyes for more durable color. With care, the natural look is preserved.

References

Al Pertuit . 2004. Drying Flowers. Website: http://hgic.clemson.edu/

Alleman , E. 1999. Here to stay from 'The Rose Ette'. Website: http://www.houstenrose.org/hrshere.htm

Anilker, R. 1981. Textileveredlurg.16:431.

Anon. 1996. Drying and processing technique for flowers. Website: http://www.msue.msu.edu/msue/imp/mod03/03900012.html

Anon . 1997 a. Is it real or freeze dried. Website: http://w3.goodnews.net/~aretz/dry.htm.

Anon . 1997 b. Identifying dynamite dried – Freeze dried. Website: http://www.safnow.org/fm/columns/june 96ws.html.

Anon . 1997c. Hydrosorbent products. Website: http://127.0.0.1:1888

Anon . 1998. Dry flower industry in India. Floriculture Today, 3(10):10-12.

Anon . 2006.Website: http://www.iihr.res.in/newsandevents/IIHR % 20 Annual % 20Report_2006-07.

Bhalla Rajesh, Moona, S.R. Dhimma and K.S. Thakur. 2006. Standardization of drying techniques of chrysanthemum (*Dendranthema grandiflorum* Tzvelev). J.Orn.Hort., 9(3): 159 -163.

Bhutani,J. C. 1995. Drying of flowers and floral craft. In: Advances of Horticulture (12): Ornamental Plants. Ed., Chadha, K. L. and S. K. Bhattacharjee. Malhotra Publishing House, New Delhi, pp.1053-1058.

Blacklock,J. 1993. In: Teach yourself dried flowers. Ed., Hodder and Stoughton., U.K.,pp.188.

Boonsong Panthip. 2008. Improvement of aroma in Thai potpourri and dry flowers by using starch film as aroma fixative. Website:http://www2.kmutt.ac.th/news/en_newsdetail.aspx?ref=200805000179

Bryan, L. 1992. In: Drying flowers and herbs. Harlexton Publishing Ltd., Linecolnshire, UK. pp. 48.

Bull,B. 1999. Drying flowers for everlasting beauty. Website: http://www.garden.rg/nga/edit/Articles/dryflowers.qua

Chen, Wei, Karen L. B. Gast and Sheri Smithey. 2000. The effects of different freeze drying processes on the moisture content, colour and physical strength of roses and carnations. Scientia Horticulturae., 84(3/4): 321-332.

Chimonidou – Pavlidou, D. 2000. New cut flowers for fresh and dry production cultivated in Cyprus. Acta Horticulturae 541: IV International Symposium on new floricultural crops (Oct 2000). Website: http://www.actahort.org/members/showpdf?booknrarnr=541_11

Chitra,R. 2001. Exploitation of under utilized flowers of pagoda tree(*Plumeria alba*) for flower drying. M.Sc. (Hort.) Thesis, Tamil Nadu Agricultural University, Coimbatore.

Conder, S. 1979. In: Commercial dried flowers. Merechurst Publication, London. pp.60-75.

Dahiya, D.S., D. Unnikrishnan, A. K. Gupta, S. K. Sehrawat and S. Siddiqui.,2003. Dehydration of Annual Chrysanthemum (*C.coronarium*). Acta Horticulturae., 624:385-388.

Dana, M.1983. Preserving plant material. Floriculture Today, 2 (1): 5-9.

Datta, S. K. ,1999. Dehydration of flowers and foliage in floral craft. Floriculture Today, 12(3): 34-35.

Doreen Fitzgerald . 2003. Peony – A Future Crop for Alaska? Agro Orealis, 35 (1): 4-11 Website: http://www.uaf.edu/snras/AFES/AFESpubs.html

Evelyn.1997. Drying flowers. Website: http://www.letshop.com/aacbb/0000053.htm

Geetha, C. K. , P. K. Rajeevan and P. K. Valsalakumari. 2004. In: Dry Flowers. Kerela Agricultural University, KAU P.O.-680 656, Thrissur, Kerela. pp:80.

Gill, Sukhmani Sher, Rupa Bakhshi and J. S. Arora. 2002. Evaluation of various methods of drying flowers and foliage. J.Orn.Hort., 5 (2): 56-57.

Gulrajani,M. L. and N. Sukumar. 1985. Optimization of a single stage preparatory process for cotton using sodium chlorite. Tex. Res. J., 56:476-483.

Gupta,V.K., R. A. Sachan, V. P. Singh, and S. K. Sharma, 1997. Natural Dyes. Indian Textile Journal, 108:16.

Irene Vethamoni P., R. Sankaranarayanan and R. K. Murali Bhaskaran. 2003. Identification and standardization of plant species suitable for dry flowers at Kodaikanal. In: One hundred research papers in floriculture A compedium of research papers of national symposium on recent advances in Indian floriculture. Rajeevan,P. K., P. K. Valsalakumari and R. L.Misra.12-14 Nov.2003, KAU, Vellanikkara, ISOH, New Delhi. pp. 350-356.

Irene Vethamoni P. and S. Nanthakumar. 2002. Standardization of technology for dry flower production at Kodaikanal. South I.Hort., 50 (4-6): 653- 657.

Jasmeet Kaur. 1999.Dry flowers for new thrust to economy. Floriulture Today, 3 (12) : 5-7.

John Van Dyk. 1998. Drying flowers. Website: http://www.ipm.iastate.edu/ipm/hortnews/dry flowers.html.

Kawasaki,M. 1975. In : Dried flower arrangements. Japan publications, INC.

Keisar Lourdasamy. D. 1998. Studies on drying, bleaching and dyeing of annual flowers and dry floral crafts. M.Sc. (Hort.) Thesis, Tamil Nadu Agricultural University, Coimbatore, Tamil Nadu.

Keisar Lourdasamy. D., E. Vadivel and R. S. Azhakiamanavalan. 2002. Studies on bleaching of annual flowers for dry flower crafts. S.Indian Hort., 50(4-6): 658-660.

Keisar,D.Lourdasamy., E.Vadivel and R.S.Azhakiamanavalan. 2003. Studies on critical stages of harvest of annual flowers for dry flower production. S.Indian Hort., 51(1-6):241-243.

Khanna,R.K. and S. Chandra. 1996. Forest / domestic waste as a source of natural dyes. J.Econ. and Taxonomic Botany, 20: 497-500.

Kher,M.A. and Bhutani,J.C. 1979. Dehydration of flowers and foliage. In: Extension Bull., EBIS, NBRI., Lucknow, p.1-20.

Macphail, B. 1997. Dried flower observational trial. Website: http://webl.nsac.ns.ca/nsdam/pt/projsum/91/91-22.htm

Mark,H. F., D. F. Othmer, C. G. Overberger and T. Seaborg. 1979. Dyes. In: Encyclopedia of chemical technology. Ed. A. Standen. Vol.3. John Wiley and Sons, USA: 151-211, 351-392.

Mckenzie, D. 1996. Growing Darwinia, grafted for reliability. Ornamental Hort. Abstr., 22(6): 2079.

Miller,K. 1999.Forever foliage preserver. Website:http// www.floralhome.com/natdry/htm

Mirunalini, P. 1999. Evolving suitable technology for dry flower production and assessment of suitable plant species. MSc. Thesis. Tamil Nadu Agricultural University, Coimbatore, Tamil Nadu.

Padamavathamma, 1999. Standardization of drying techniques for Statice dry flower. MSc. Thesis, University of Agricultural Sciences, Bangalore, Karnataka.

Paparozzi, Ellen, T and Diane E. McCallister,. 1988. Glycerol and microwave preservation of annual Statice (*Limonium sinuatum* Mill.). Scientia Hort., 34 (3/4): 293-299.

Priyesh,S. , C. K. Geetha, P. K. Rajeevan, P. K. Valsalakumari and Shalini Manoharan. 2003. Post harvest treatments for value addition in celosia. In : One Hundred Research Papers In Floriculture. A Compedium of research papers of National Symposium On Recent Advances In Indian Floriculture. Rajeevan, P. K., P. K. Valsalakumari and R. L. Misra. 12-14 Nov.2003, KAU, Vellanikkara, ISOH, New Delhi.pp. 357-359.

Raj, Desh and P. Gupta. 2005. Standardizing dehydration technology for ornamental herbaceous plant from outer Himalayas. J. Orn. Hort., 8 (1): 53-55.

Rebecca Gonzalez, Nahid Abdelnour and Mariana Gil. 1995. Drying Flowers. Science 4(1): 8-9 http://youth.net/nsrc/nsrc.html

Rothenberger, R. R. 1997. Drying flowers and foliage for arrangements. Agricultural Publication GO 6540. Dept. Hort. University of Missouri. Columbia. Website: http:// www.muextension.missouri.edu/xplor/agguides/hort/go6540.htm

Saffena,S. A., V. S. Patil and B. Hemla Naik. 2006. Response of drying in hot air oven on quality of rose flowers .J.Orn.Hort., 9 (2): 114-117.

Samant,Singh, P. K. and Dilip Kumar Dash.1998. Studies on the effect of dessicants on drying of roses. Orissa J.Hort., 26(1):88-90.

Sangama. 2003. Studies on dehydration of Gomphrena globosa flowers. In: One Hundred Research Papers In Floriculture. A Compedium of research papers of National Symposium on Recent Advances In Indian Floriculture. Rajeevan,P. K. , P. K. Valsalakumari and R. L. Misra.12-14 Nov.2003, KAU, Vellanikkara, ISOH, New Delhi. pp. 360-363.

Sangama. 2004. Dehydration and product diversification of Helichrysum flowers. J.Orn. Hort., 7(3-4): 376 -380.

Seaberg, P. 1996. Drying Roses. Website: http://www.ars.org/drying.html

Sell,Randy and Dwight Aakre. 1993. Dried and fresh cut flowers. Alternative AgricultureSeries.No.13.Website: http://www.ag.ndsu.edu/pubs/alt-ag/flowers.html

Semant, P. K. S., Dash, D. K., Barik, L. R. and Dhar, A. N. 1993. Effect of glycerine on preservation of foliage of some ornamental plants. Orissa J.Hort., 21(1&2):73- 75.

Singh, Alka and B. K. Dhaduk. 2005. Effect of dehydration techniques in some selected flowers. J.Orn.Hort., 8 (2): 155-156.

Singh, Alka, B. K. Dhaduk. and R. K. Shah. 2003. Effect of dehydration on post harvest life and quality of zinnia flowers. J.Orn.Hort., 6 (2): 141-142.

Song Chae Eun, Kim KwangSoo, Joang HongGi, Yoo YongKweon, Heo BukGu and Park Yun J Um. 2004. Natural Dye on dry flower, the *Lagurus ovatus* using the extract isolated from marigold. Korean J. Science & Tech., 22(1):119-124.

Stewart, M. 1997. Drying roses. Website: http://www.toreadors.com/martha/projects/roses.html

Sujatha, A. N. , Damodaran, T. and Shiva, K. N. 2001. Dry flower industry in Andamans. Kisan World, 28:28.

Thomas, A. Jennings.1999. Lyophilization: Introduction and Basic Principles. Forma Health Care., pp12-13.

Thomler, J. 1997. Drying flowers and leaves. Website: http://www.nectar.com.au./jascraig/craft/drieddf.htm

Tom Harris. 1999. How Freeze- Drying works. Website: http://recipes.howstuffworks.com/freeze-drying.htm

Trinklein, David. 2006 Storing dried material. extension. missouri. edu. Website: http://www.muextension.missouri.edu/explore/agguides/hort/go6540.htm

Venkataraj ,R. and Gulrajani, M. L. 1989. Bleaching of polyester/cotton blended fabrics with a formed composition. Ind.J.Fib. and Tex.Res.14 (3): 35-38.

Venugopal,C. K. and A. A. Patil. 2000. A note on the effect of drying methods and temperatures on colour intensity of everlasting flower. Karnataka J. agric. Sci.,13(3) : 793-794.

Verey, R. 1998. The flower arrangers garden. Drying flowers. Personal communication E-mail: twep-webmaster@pathfinder.com

Westland, Pamela. 1992. The complete flower arranger. Anness Publishing Ltd., London. pp.256.

Wilkins,H. F. and S. L. Desborough., 1986. Cryo -Drying of *Dianthus Caryophyllus* L. flowers. Acta Horticulturae, 181:477-481.

William,C. and Segal.1972. Dyes and Dyeing Processes, In : Encyclopedia of Textiles. Ind. Ed. Doric Publishing Company, USA.pp 439-449.

Yockley,T. L. 1997. Everlastings, Website: http://www.geocities.com/Rainforest/1329/everlast.htm

Yogita Chari 2000. Studies on the standardization of bleaching and dyeing technology of flowers. MSc. Thesis., Tamil Nadu Agricultural University, Coimbatore, Tamil Nadu.

Reference

- **National Gardening Association**
  NGA was found in 1973 and is a non-profit organization, leader in plant based education.
  www.garden.org
  National Gardening Association, Dorset Street, South Burlington, VT 05403
  Phone: 802-863-5251

- **Association of Specialty Cut Flower Growers, Ohio**
  The ASCFG is a well-run organization, which addresses all aspects of the floral business.
  www.ascfg.org

- **British Dried Flower Association**
  The British Dried Flowers Association is a Trade Body of flower growers, specializing in the production and marketing of dried flowers, hops, herbs and grasses.
  www.flowergrowers.co.uk
  Stonedge, Manor Road, Staverton, Nr. Daventry
  Northants NN11 6JD

- **The Spice Association of Australia**

  The Spice Association of Australia was formed in 1987 to provide an industry voice for those trading in dehydrated vegetables, herbs and spices.

  PO Box 104, St Leonards, NSW 2065, Australia

- **Dried Flowers Galore**

  It is a growing community of dried flower enthusiasts, novice to expert, who share their experiences, knowledge and all their drying expertise.

  www.dried-flowers-galore.com

Dried Flower Books and Literature

1. Agnew, Nancy H. and Naeve, Linda. "Growing and Drying Everlasting Flowers," PM1398, Ames: Iowa State University Extension Service, July 1990.
2. Bird, Lindy. "Dried Flowers: A Complete Guide, Amazon.co.uk.
3. Gertley, Jan and Gertley, Michael. "Dried Flowers for All Seasons: Creating the Fresh-Flower Look Year-Round", Amazon.com.
4. Miller, Cathy and Gray, Rob. "Harvesting, Preserving & Arranging Dried Flowers", Amazon.com.
5. Olson, Beverly and Lazzara, Judy. Country Flower Drying, New York: Sterling Publishing Company, 1988.
6. Smith, Ronald C. "Methods of Preserving Flowers," Fargo: North Dakota State University Extension Service, April 1992.
7. Sterbenz, Carol Endler and Felber, Richard. "The Complete Book of Dried Flower Topiaries: A Step-By-Step Guide to Creating 25 Stunning Arrangements", Amazon.com.
8. Patricia Thorpe, Everlastings, the Complete Book of Dried Flowers, ISBN 0-395-41160-2, Houghton Mifflin Company, Quarto Marketing Ltd, 15 West 26th Street, New York, NY 10010, 1985.

Publications

1. **Freeze Drying Product**

   Analysis of dried product for process development

   www.lyophilizationtechnology.com

2. **Decas Dried Cranberries**

   Wholesale quality dried cranberries for trail mixes and nutritional bar www.decascranberry.com

3. **A Year Full of Flowers: Fresh Ideas to Bring Flowers Into Your Life Everyday**
   Jim McCann, Julie McCann-Mulligan
   www.bestprices.com

4. **Dried Flowers: A Complete Guide**
   Lindy Bird
   www.buy.com

5. **Microwave Pressed Flowers: New Techniques for Brilliant Pressed Flowers**
   Joanna Sheen
   www.walmart.com

6. **Dried Flowers for All Seasons: Creating the Fresh-Flower Look Year-Round**
   Gertley, Jan and Gertley, Michael
   www.amazon.com

7. **The Complete Book of Dried Flower Topiaries: A Step-By-Step Guide to Creating 25 Stunning Arrangements**
   Sterbenz, Carol Endler and Felber, Richard.
   www.amazon.com

8. **Harvesting, Preserving & Arranging Dried Flowers**
   Cathy Miller, Rob Gray
   www.amazon.com

Magazines

1. Fusion Flowers
   www.fusionflowers.com/ welcome.htm

2. Flower Arranger
   www.amazon.com

3. Flourists Review
   www.amazon.com

4. Floralartmall.com
   www.floralartmall.com/ magazines.html

Websites

1. www.FlowersToLebanon.com

2. www.therightstrains.co.uk

3. www.greenmonsterIncense.com
4. www.o-hana.jp/
5. www.swissflowers.ch
6. www.leafshine.eu
7. www.merk-mdp.com
8. http://www.poppiesshop.com/
9. http://www.dry-flowers.net/
10. http://pages.prodigy.net/hansy/pampas.htm
11. http://www.eindiabusiness.com/indianpotpourri/
12. http://www.naturalsbyweb.com/
13. http://www.piercearnold.co.uk/
14. http://www.woodcreekdrieds.com/
15. http://www.fourseasonsflowers.com/
16. http://www.kooslamboo.com/
17. http://www.poppies.ws/
18. http://www.rbdriedflowers.com/
19. http://www.cheappoppies.com/

Glossary

1. **Absorbent -** Material capable of absorbing and holding liquids by the capillarity of the pores between or within the closely matted cellulosic fibers.

2. **Accent -** Accent may also accomplished by introducing sharp contrast in form, size or color. Arrangements should have a focal point to attract immediate attention or interest.

3. **Air Drying -** Commonly referred to as the "hang and dry" method. The stems of flowers and their foliage are tied and hung upside down and dried.

4. **Alum -** Alum is used to clarify water during dye extraction.

5. **Annuals -** Plants that complete their life cycle within the space of one season or one year.

6. **Artificial Flowers -** Great source of handmade decorations. Beautiful handmade artificial flowers are suitable for bouquets, weddings or to decorate any room in the home. They are generally made of paper or cloth and in a variety of colors.

7. **Asymmetrical Balance** (informal balance) **-** The plant material and placement are different on each side of the central axis.

8. **Balance -** Balance is the one aspect that gives a floral arrangement its visual appeal and determines whether it can be classified as a formal or an informal arrangement. Balance is related to the correct use of color and weight of flowers (the visual weight and not its actual weight). Generally dark and dense flowers appear heavy. As a rule, for a balanced appearance, heavy flowers are placed at the bottom of the arrangement and lighter ones towards the outer edges and upper portions.
9. **Basket Bouquet -** A beautiful one-sided arrangement of dry flowers that add aesthetic beauty and charm to the décor of a house or an office.
10. **Borax –** is a dessicant used for drying flowers.
11. **Box Bouquet -** Flowers are well represented in the complete spectrum of fine, durable boxes. A large array of flowers can be arranged in a box giving a natural look to the flowers.
12. **Boxes** (dry flower decorated boxes) **-** are made from a wide range of materials with innovative, attractive designs, patterns made from dried flowers and leaves.
13. **Caking -** A layer or deposit of compacted matter
14. **Candle Holders -** Using dried flowers, candle holders are decorated.
15. **Cane Coils -** Long canes with a coiled structure at one end
16. **Cardboard Box -** A paper product made of unbleached craft paper, with two heavy outer layers and inner layer to provide strength. Boxes are made of cardboard for packing of dry flowers.
17. **Carotenoids –** Carotenoids are yellow, orange and red pigments synthesized by plants.
18. **Carry Bags -** are of multipurpose uses. Jute and dry flowers décor together form a completely eco-friendly bag. Using dried flowers, leaves and foliage a number of designs and patterns are made on the jute bag. They can also be customized by a special message written by dried leaves. These form wonderful gifts.
19. **Collages -** A paste-up made by sticking together pieces of pressed and dried flowers and foliage.
20. **Condensing -** To undergo condensation, to become more compact
21. **Cones -** is a general term used for the fruit of a pine tree. A pine cone is shaped like an egg or a cone. It has scales on them. Pine cones usually have spiny points at the tip of each scale. There are a number of species falling under the pine category.

22. **Contrast -** Stimulates interest. It is achieved by placing opposite or unlike elements together so as to emphasize or accentuate their differences.

23. **Cryo Drying -** is drying under low temperature.

24. **Crystal Potpourri -** A unique gel like crystal that can be simmered or left out to fragrance a room.

25. **Sachet -** A small packet of perfumed powder used to scent clothes, as in trunks and closets.

26. **Crystallizing -** To cause to form crystals or assume a crystalline structure.

27. **Desiccant -** A substance with a high affinity for water which can be used as a drying agent.

28. **Dominance -** Dominance in floral arrangement is the greater force of a design element, such as more round forms, or more curved lines. Dominance may be achieved through the use of a dominant color, size, shape or texture, or by using larger forms or stronger colors.

29. **Dried Flower Bags -** Elegant pieces of dried flower bags are available today in the worldwide market. These are offered in a variety of colors and features beautiful designs are made by dried flowers.

30. **Dried Leaves -** Leaves are dried or preserved to capture the beauty of spring and hold it for a long time.

31. **Dried Petals -** Many types of dried petals are used for making potpourii eg., land lotus petals, rose petals, mahogany petals, lily petals and lotus petals.

32. **Dried Pods / Fruits -** The enchanting beauty and usage of fresh pods or fruits need not end up with the season. They are dried and preserved for years. A colorful array of choice dried fruits are very popular in dried arrangements. There are numerous dried fruits and pods used in potpourri and dried flower arrangements. There is a wide variety of fruits that are successfully used in their dried form and are dried by different techniques depending on their types and requirements.

33. **Driftwood -** A wood that has been washed on to a shore or beach of a sea or river by the action of winds, tides, waves or man. It is a form of marine debris or tide wrack.

34. **Dry Flower Cards -** The greeting cards made using dried flowers.

35. **Dry Flower Sachet -** A sachet is a small packet of perfumed powder used to scent clothes, as in trunks and closets. By using dried flowers, a memorable gift can be made in the form of dry flower sachet.

36. **Dry Flower Wreath -** A wreath is a ring or circlet of flowers, boughs, or leaves worn on the head, placed on a memorial or hung as a decoration. By using dried flowers, a colorful wreath can be created to hang on the front door or over the mantel.

37. **Embedding -** The process by which flowers are immersed in dessicants for drying.

38. **Emphasis -** Emphasis is used to first attract attention to the most important feature in an arrangement (focal point) and from that point to every detail in order of its importance.

39. **Enzyme Aided Extraction -** Enzymes are used to facilitate the extraction, increase pigment yield and improve the stability of the resultant colour preparation.

40. **EXIM Policy -** Export Import policy.

41. **Floral Arrangement -** The art of selecting and organizing flowers and foliage according to the elements and principles of design in order to attain a pleasing and interesting design.

42. **Floral Foams -** are available in wet or dry forms; wet foams are used for arranging fresh plant materials, while dry floral foam is used for dried or silk floral arrangements. Wet and dry types of floral foams are not interchangeable – their intended use is specific.

43. **Flower Press -** Two pieces of corrugated cardboard, newspaper, a few sheets of blotter or printer paper, and three large rubber bands are required to use this press. The flowers should be placed between a few pieces of paper, between the pages of the newspaper, the newspaper is sandwiched between the cardboard, and held together with the rubber bands, evenly spaced. Pressure is applied by placing heavy objects on top.

44. **Focal Point -** That area or portion of a design which first attracts attention is the focal point of an arrangement

45. **Form -** It describes the three-dimensional aspect of a design in flower arrangement. A form may be closed – compact, massed, with few open spaces which is typical of a mass design.

46. **Garland -** A wreath or rope of flowers and leaves.

47. **Glycerin Drying -** In this method, moisture in a flower or foliage is replaced with glycerin and water.

48. **Handmade Paper** (for dried decorations) - In making handmade paper, no chemicals and harmful dyes are used. It is made by using old cotton rags, dried flowers and other materials.

49. **Harmony** - Is an attribute of design in flower arrangement and is the principle that produces a feeling of unity throughout the entire composition.

50. **Herbarium** – A collection of press dried flowers and foliage.

51. **High Density Polyethylene** - It is made from a thinner gauge material to lower freight costs and storage requirements. It has great strength and is water-resistant.

52. **Holders** - A good holder should give the freedom to position stems where they are wanted and hold them in place securely.

53. **Hydrogen Peroxide Bleaching** – Bleaching agent for dry flowers and foliage. The cleavage of double bonds breaks up the coloured molecules and bleaching takes place.

54. **Lamp Shades** - Dried flowers are used to decorate lamp shades. These flowers look beautiful when light shining through them.

55. **Low Density Polyethylene** - It is the material used in most poly bags. It has greater tensile strength, clarity, flexibility and water-resistant properties.

56. **Lichens** - are composite organisms consisting of a symbiotic association of a fungus (the mycobiont) with a photosynthetic partner (the photobiont or phycobiont), usually either a green alga (commonly Trebouxia) or cyanobacterium (commonly Nostoc). Lichens have also been used in making dyes and perfumes, as well as in dry flower decoration.

57. **Lyophilization** – is the process of lowering the temperature of an object and then using a vacuum to extract all the moisture from the item.

58. **Microwave Oven Drying** - This is drying done by microwave oven. It is quick and relatively simple.

59. **Mordants** - Mordants are substances which are used to fix dyes to the material.

60. **Moss** - Mosses are small, soft plants commonly growing close together in clumps or mats in damp or shady locations. There are approximately 12,000 species of moss classified in the Bryophyta (Division of Plantae Kingdom).

61. **Pomanders -** A pomander is a ball made of perfumes. The term "pomander" can be for the actual scented material itself or for the container that contains the scented material. Pomander can be a bag containing fragrant dry flower material.
62. **Natural Drying –** It involves sun drying in the field, requiring large amount of space.
63. **Natural Dyes -** Dyes extracted from plants, animals and minerals.
64. **Oshibana -** The art and craft of pressing flowers to retain the benefits and beauty of the living plant.
65. **Palm Caps -** Open pods of the palm trees
66. Pattern is the design formed by solids and spaces. Individual components in a design have a pattern of their own; individual components are combined into an overall pattern.
67. **Perennials –** Plants which have a life span of more than one year and usually bloom during a specific time period every year during the growing season. These flowers are hardy and herbaceous.
68. **Photo Frames -** Different colors and kinds of dried flowers are used to beautify the photo frames.
69. **Pinpoint Holders -** These needle holders are usually round, though ovals are also available. They range in size from 1 to 6 inches across. Protruding from the base are sharp, heavy-duty metal prongs, which are spaced closely together to form a tight mass.
70. **Poly Bag -** A bag that is made of polyethylene, which is a flexible, water resistant plastic that is ideal for making merchandise bags.
71. **Potpourri -** A mixture of dried, sweet-scented plant parts including flowers, leaves, seeds, stems and roots. This is a special type of dry flower arrangement which is made as a decorative piece and a room freshener.
72. **Pounding -** The act of beating, bruising or breaking up of raw material to extract natural dyes.
73. **Press Drying -** One of the most popular methods for drying flowers. Flowers are pressed and used in making greeting cards etc.
74. **Proportion -** Important in the use of color, texture or form. Components in the arrangement should be in good proportion to one another.
75. **Quilt -** Wall quilts made of dried flowers introduce a little of summer hues in the home and add zest to the living room. They are available with many botanical prints and its fun to make these quilts.

76. **Rhythm** - Refers to movement. The swing or motion in an arrangement should create a dominant visual path.

77. **Selaginellas** - are creeping or ascendant plants with simple, scale-like leaves on branching stems from which roots also arise. It is a genus of plants in the family Selaginellaceae.

78. **Scale** - Size relationship between flowers or between flower and container.

79. **Scrape Book** - An album into which clippings or notes or pictures can be pasted.

80. ***Silica* Gel** – A desiccant used for drying flowers

81. **Skeleton Leaves** - Process of removing chlorophyll from leaves for dry flower making.

82. **Skimming** – To remove floating matter from liquid

83. **Soaking** – Raw material immersed in liquid over night to facilitate extraction of natural dyes.

84. **Sodium Chlorite Bleaching** - This bleach removed the entire colour from cellulose based material with minimal damage, even during prolonged contact.

85. **Sodium Hypochlorite Bleaching** - The hypochlorites are one of the strongest classes of bleaching agents known.

86. **Sola Flowers** - Handmade flowers from tapioca wood.

87. **Solar Drying** – It is a method for drying plant materials using solar energy.

88. **Stem Supports** - Many flower heads are too heavy for the dried stems to support during the drying process. The stems have to be cut and wired before hanging them out to dry.

89. **Sublimation** - It is the principle based on which freeze drying is done. In this process, water is removed as vapor directly from ice, without passing through the liquid state.

90. **Supercritical Carbon-di-oxide Extraction** - The use of super critical carbon di oxide as an extracting solvent is one of the modern approaches to avoid the detrimental effect of heat and light and also to reduce the interference of co-extracted material.

91. **Swags** - Green or dried material secured to some sort of flat frame. Dried floral swags can enhance a doorway or banister for holidays and events.

92. **Symmetrical Balance** (formal balance) - Symmetrical balance requires a repetition of flowers, buds, and foliage on both sides of the central axis with the two sides as nearly alike as possible.

93. **Texture** - Plant texture refers to the surface quality of plant material. Textural contrast and variety can add interest to a design.

94. **Topiary** - They have a variety of dried flowers arranged to take on the shapes of animals, lettering, numbers, or various but precise geometric forms.

95. **Traditional Fermentation Process** - This method is used for extraction of indigo, annatto etc. The process involves the disruption of plant material followed by acidified aqueous extraction for chlorophyll, carotenoids and annatto.

96. **Unity** - An attribute of design in flower arrangements, is that quality that expresses congeniality, cooperation, and a spirit of working together as a unit.

97. **Vacuum Drying** - In this method, heat is supplied by conduction or radiation (or microwaves) while the vapor thus produced is removed by the vacuum system.

98. **Visual Balance** - Visual balance depends upon placement of flowers and foliage. Visual balance makes the appearance of a floral arrangement balanced to the eye.

99. **Water Drying** - Water drying is a method of preserving in which the leaves are stripped off and the flower stems placed in five centimeters of water.

100. **Waxed Paper** - Paper that has been impregnated with wax to make it waterproof or greaseproof. It is used while colouring ferns and small leaves.

101. **Window Boxes** - are gardens in a box, manageable and enchanting. Various types and colors of dry flowers are beautifully arranged in the window boxes.

Annexure

Table 1 : Plants and Their Parts Suitable for Collecting and Drying

Flower or leaf thickness	Minimum times
Silica gel	
Thin textures	2 days
Medium textures	3-4 days
Heavy textures	5-7 days
Other dessicants	
Thin textures	4-5 days
Medium textures	6-9 days
Heavy textures	10-14 days

Table 2 : Suggested cooking and standing times for specific flowers

Flower	Cooking time	Standing time
Roses	2 1/2 min.	overnight
Daisy-type flowers: zinnias, marigolds, daisies, chrysanthemums	1 1/2 min	10 hours
Carnations	1 1/2 min.	10 hours
Large dahlias	3 min.	36 hours

Large chrysanthemums	3 min.	36 hours
Peonies	3 min	36 hours
Small orchids	1 1/2 min	24 hours
Large orchids	2 1/2 min	24 hours

Table 3 : Suggested drying methods for foliage

Scientific Name	Common Name	Air Dry	Glycerine Treat	Press	Skeletonize
Acer	maples	-	X	X	X
Agave	century plants	X	X	-	-
Alpinia	shell ginger	X	X	-	-
Araucaria	monkey puzzle tree	X	X	-	-
Artemesia	wormwood	X	-	X	-
Asclepias	milkweed	X	-	X	-
Aspidistra	cast iron plant	X	X	-	-
Bambusa	bamboo	X	X	-	-
Berberis	barberry	X	X	-	-
Bromeliaceae	bromeliads	X	-	-	-
Buxus	boxwood	X	-	-	-
Caladium	caladium	-	-	X	-
Callistemon	bottlebrush	-	-		-
Carya	hickory	-	X	X	X
Cecropia	cecropia	-	-	X	-
Coccoloba	sea grape	X	-	X	-
Cocculus	snail seed	X	-		-
Cordyline	dracaena, ti	X	-		-
Crataegus	hawthorn	-	X		-
Croton	croton	-	-		-
Cycadaceae	cycads	-	-	X	-
Cyperus	papyrus	-	-		-
Eriobotrya	loquat	X	-		-
Eucalyptus	silver dollar	X	X		-
Fatshedera	fatshedera	-	-	X	-
Fatsia	fatsia	-	-	X	-
Ferns	many genera and species	X	X	X	-

Ficus	figs	-	X	-	X
Hamamelis	witch hazel	X	X	-	-
Hedera	ivy	X	X	X	-
Heliconia	heliconia	X	X		-
Herbs	many genera and species	X	-	X	-
Hosta	plantain lily	X	-	X	-
Ilex	holly	-	X	-	-
Illicium	anise	X	X	-	-
Juniperus	juniper, cedar	X	X	-	-
Lichens	lichens	X	-	-	-
Ligustrum	privet	-	-	-	-
Lycopodium	club moss	-	-	-	-
Magnolia	magnolia	-	-	-	X
Mahonia	grape holly, Oregon grape	-	X	X	-
Melaleuca	punk tree	X	X	-	-
Myrica	myrtle	X	X	-	-
Palmaceae	palms	X	X	-	-
Pandanus	screw pine, southern yew	X	-	-	-
Podocarpus	podocarpus	X	-	-	-
Prunus	plum and cherries	-	-	-	-
Quercus	oaks	-	-	-	X
Rumex	dock	X	-	-	-
Russelia	firecracker plant	X	-		-
Selaginella	selaginella	X	X	X	-
Taxodium	cypress	X	X		-
Taxus	yew	X	X		-
Tetrapanax	rice paper plant	-	-	X	-
Thuja	arborvitae	X	X		-
Trevesia	snowflake plant	-	-	X	-
Vaccinium	huckleberry	X	-	-	-
Viburnum	viburnum	-	X	-	-
Yucca	yucca	-	X	-	-
Zamia	coontie	X	X	-	-

Table 4 : Suggested drying methods for flowers

Scientific Name	Common Name	Natural Dry	Air Dry	Desiccate	Press
Acalypha	chenille plant	-	X	-	-
Achillea	yarrow	X	X	-	X
Agapanthus	lily of the Nile	-	-	X	-
Ageratum	floss flower	-	X	X	-
Alcea	hollyhock	-	X	X	-
Allamanda	golden trumpet	-	-	X	-
Allium	onion	-	-	X	-
Aloe	aloe	-	-	X	-
Alstroemeria	lily of the Incas	-	-	X	-
Althaea	mallow	-	X	X	-
Amaranthus	amaranth	X	X	-	-
Anthemis	marguerite	-	-	X	X
Anthurium	flamingo flower	X	X	-	-
Antirrhinum	snapdragon	-	-	X	-
Aphelandra	zebra plant	-	-	X	-
Arctotis	blue eyed daisy	-	-	X	-
Artemesia	wormwood	X	-	-	X
Astilbe	spirea	-	X	-	X
Belamcanda	blackberry lily	-	-	X	-
Calendula	calendula	-	-	X	-
Calla	calla	-	-	X	-
Calycanthus	sweet shrub	-	-	X	-
Camellia	camellia	-	-	X	-
Celosia	princess plume and cockscomb	-	X	X	-
Chrysanthemum	mum, daisy, feverfew etc.	X	X	X	X
Cirsium	thistle	-	X	-	-
Clematis	virgin's bower	-	-	X	X
Compositae	daisy and daisy-like flowers	-	X	X	X
Consolida	larkspur	-	-	X	-
Cornus	dogwood	-	-	X	-
Cosmos	cosmos	-	-	X	X
Crinum	crinum lily	-	-	X	-
Crossandra	firecracker plant	-	-	X	-

Cynara	artichoke, cardoon	-	X	-	-
Dahlia	dahlia	-	-	X	-
Delphinium	delphinium	-	-	X	-
Dianthus	pink, sweet William, carnation	-	-	X	X
Echinops	globe thistle	-	-	X	X
Erica	heather	X	X	-	-
Eupatorium	boneset	-	X	-	X
Gaillardia	blanket flower	-	-	X	X
Galphimia	thryallis	-	-	X	-
Geranium	geranium	-	-	X	-
Gerbera	African daisy	-	-	X	X
Gladiolus	gladiolus	-	-	X	-
Gomphrena	globe amaranth	X	X	-	-
Grevillea	silk oak	-	-	X	-
Gypsophila	baby's breath	X	X	-	X
Helichrysum	straw flower	X	X	-	-
Heliconia	lobster claw	X	X	X	-
Hibiscus	hibiscus	-	-	X	-
Hippeastrum	amaryllis	-	-	X	-
Hydrangea	hydrangea	X	X	-	-
Iris	iris, flag	-	-	X	-
Ixora	ixora	-	-	X	-
Justicia species	jacobinia, Brazilian plume, shrimp plant	-	-	X	-
Liatris	liatris	-	X	X	-
Limonium	statice	X	X	-	-
Mathiola	stock	-	-	X	-
Molucella	bells of Ireland	-	X	X	-
Narcissus	daffodil	-	-	X	-
Orchidaceae	cattleya, cymbidium, etc.	-	-	X	-
Passiflora	passion vine	-	-	X	-
Penstemom	beard tongue	-	-	X	-
Pentas	star cluster	-	-	X	-
Protea	protea	-	X	X	-
Pyrostegia	flame vine	-	-	X	-
Reseda	mignonette	-	-	X	X
Rhododendron	azalea	-	-	X	-

Rosa	rose	-	-	X	X
Rudbeckia	blackeyed Susan	X	X	-	-
Rumex	dock	X	X	-	-
Russelia	firecracker plant	-	-	X	-
Salvia	sage, salvias	-	-	X	-
Sarracenia	pitcher plants	-	X	X	-
Senecio	groundsel	-	-	X	X
Solidago	golden rod	X	X	-	X
Spathodea	African tulip tree	-	-	X	-
Spathiphyllum	spathe flower	-	X	X	-
Strelitzia	bird of paradise	-	X	X	-
Stokesia	stokes aster	-	-	X	X
Tagetes	marigold	-	-	X	-
Tritonia	montbretia	-	-	X	X
Verbena	verbena	-	-	X	X
Viola	pansy, violet	-	-	X	X
Zingiber	ginger, pine cone lily	-	X	X	-
Zinnia	zinnia	-	-	X	-

Table 5 : Suggested drying methods for fruits

Scientific Name	Common Name	Natural Dry	Air Dry
Abelmoschus	Okra	X	X
Acacia	mimosa, acacia	X	X
Acer	maple	X	X
Aesculus	horse chestnut	X	X
Agapanthus	lily of the Nile	X	X
Agave	century plant	X	X
Althaea	hollyhock, mallow	X	X
Araucaria	monkey puzzle tree, Norfolk Island pine	X	X
Asclepias	milkweed	X	X
Bixa	lipstick tree	X	X
Blighia	akee	-	X
Bombax	red silk cotton tree	X	X
Bromeliaceae	most species (Air Plants)	X	X
Bucida	black olive	X	X
Caesalpinia	poinciana	-	X
Callistemon	bottlebrush	X	X

Campsis	trumpet vines	X	X
Capsicum	peppers	-	X
Cathamus	safflower	-	X
Carya	hickory	X	X
Catalpa	Indian bean	X	X
Ceiba	kapok	-	X
Cinnamomum	camphor	-	X
Cirsium	thistle	X	X
Clerodendrum	glory bower	-	X
Clusia	Scotch attorney	-	X
Clytostoma	trumpet vine	-	X
Cochlospermum	silk-cotton	-	X
Cocos	coconut (fruit calyx)	X	-
Combretum	combretum	-	X
Crescentia	calabash	X	X
Cycadaceae	cycads	X	X
Cynara	artichoke	-	X
Dalbergia	sissoo	-	X
Datura	angels trumpet	-	X
Diospyros	persimmon	-	X
Dipsacus	teasel	X	X
Dombeya	dombeya	-	X
Enterolobium	ear tree	X	X
Erythrina	coral tree	-	X
Eucalyptus	gum tree	X	X
Euonymus	spindle tree	-	X
Ficus	figs	-	X
Glycine	soybeans	-	X
Gossypium	cotton (calyx)	X	X
Gourds	many types	X	X
Graminae	grasses (most species)	X	X
Heliconia	lobster claw	-	X
Herbs	dill, anise, etc.	X	X
Hibiscus	hibiscus, mallow	X	X
Hippeastrum	amaryllis	X	X
Hydrangea	snow ball, hydrangea	X	X
Illicium	anise	X	X
Iris	flag, iris	X	X

Jacaranda	jacaranda	X	X
Kigelia	sausage tree	X	-
Lagerstroemia	crepe myrtle	X	X
Lilium	lily (most)	X	X
Liquidambar	sweet gum	X	-
Litchi	lichi	X	X
Lunaria	honesty	X	X
Lycium	peppers	X	X
Macadamia	macadamia nut	X	X
Magnolia	magnolia	X	X
Melaleuca	honey myrtle	X	X
Molucella	bells of Ireland	X	X
Nelumbo	lotus	X	X
Nigella	love-in-a-mist	-	X
Orchidaceae	orchids (most)	-	X
Pachira	shaving brush tree	-	X
Palmae	palms (most)	X	X
Pandanus	screw pine	-	X
Pandorea	bower plant	-	X
Papaver	poppies	X	X
Parkinsonia	Jerusalem thorn	-	X
Paulownia	princess tree	-	X
Physalis	Chinese lantern	X	X
Picea	spruce	X	X
Pinus	pine	X	-
Pittosporum	pittosporum	X	X
Porana	snow creeper	-	X
Probiscidia	unicorn plant	X	X
Protea	protea	X	X
Punica	pomegranate	X	X
Pyracantha	firethorn	-	X
Quercus	oaks	X	X
Raphanus	radish	X	X
Rhapiolepis	Indian hawthorn	-	X
Rhodomyrtus	downy myrtle	-	X
Rumex	dock	X	X
Russelia	firecracker plant	-	X
Rhus	sumac	X	X

Samaneae	monkey pod tree	X	X
Solanum	many species	-	X
Sorghum	sorghum	X	X
Spathodea	African tulip tree	X	X
Spathiphyllum	spathe flower	-	X
Sterculia	sterculia	-	X
Swietenia	mahogany	-	X
Tabebuia	trumpet tree	X	X
Tectona	teak	-	X
Tsuga	hemlock	X	-
Typha	cattail	X	X
Yucca	yucca, Spanish bayonet	X	X
Zamia	coontie, sago palm	X	X

X - indicate of drying most suitable method

FSC
www.fsc.org
MIX
Papier | Fördert gute Waldnutzung
FSC® C083411

Zeitfracht Medien GmbH
Ferdinand-Jühlke-Straße 7
99095 Erfurt, Deutschland
produktsicherheit@kolibri360.de